Forests for Whom?

Destruction & Restoration in the U.P. Himalayas

Forests for Whom?

Destruction & Restoration in the U.P. Himalayas

Neeru Nanda

HAR-ANAND
PUBLICATIONS PVT LTD

HAR-ANAND PUBLICATIONS PVT LTD
364-A, Chirag Delhi, New Delhi-110 017

PRINTED IN INDIA

Published by Ashok Gosain and Ashish Gosain for Har-Anand Publications Pvt Ltd and printed at Exel Prints, New Delhi.

This book is dedicated to

my beloved parents

and

the loving memory of

Kunta Aunty

who gave me a home,

away from home, in Dehra Dun

Acknowledgements

Senior service colleagues M.N. Buch IAS and Dr. Bhupinder Singh IAS were the prime movers for my research project which was made possible through a small grant sanctioned by the Planning Commision. Logistic and other support was extended by the National Centre for Human Settlements and Environment, Bhopal of which M.N. Buch is the Chairman. Sarla Sahney IAS, who is no more in this world, introduced me to key departments in the State Secretariat in UP and helped me get much of the information at the state level. Tsering Yongzam, also of the IAS, welcomed me to her home in Tehri and gave invaluable logistic support. Deputy Conservator of Forests STS Lepcha, IFS, and C.S. Nainwal DFO Civil Soyam took me on extensive field tours of nurseries and plantations. STS Lepcha in particular, conducted the first documentation of people's protected forests through his field staff. My debt to him is very special. In Dehra Dun, during my 16 months stint of study leave, Dr. M.R. Kohli, Mrs Kohli, Col and Mrs V.N. Mathur provided affection and sustenance. Mrs. Kohli or Kunta Aunty as we knew her, set me up my cottage in Dehra Dun and fed and nursed me through various illnesses. Today she is no more with us.

Later, during the intervening decade, Kusum Rawat has been my constant companion, in Tehri bearing my ill temper and constant goading which has finally resulted in a magnificent documentation of people's protected forests, largely managed by small women's groups all over Tehri Garhwal, completed under her leadership as District Coordinator of Mahila Samakhya of Tehri Garhwal.

Anil Agarwal, Director of the Centre for Science and Environment, New Delhi, read my earliest papers with appreciation and also published some of them. Rukun Adwani of Oxford University Press, New Delhi, read the first rambling version of this book with exemplary patience and though I could not follow his advice, his kind words of encouragement sustained me over the years while I was writing and rewriting this book. My sister Reena read the first part of the book with interest and gave useful comments.

This book woul never have seen the light of day had it not been for Narendra Kumar who published my last book and gave me the faith to persevere on this one. And to Fayaz, who worked diligently on the manuscript, taught me to use the computer and goaded me through the last difficult days when both the flesh and spirit were weak and unwilling to proceed further-a debt difficult to repay.

Contents

Introduction

Work on this book started in 1986 when I took a two year sabbatical from the Ministry of Home Affairs, Government of India, to study afforestation strategies in the catchment of the Bhagirathi river. The Bhagirathi originates from the Gangotri glacier in Uttarkashi district of the central Himalaya and is joined by river systems along the way before it emerges from the Himalayan foothills at Haridwar where it becomes the famous and holy Ganga revered by Hindus all over the world.

The Himalayas have always fascinated me. I started my career in the Indian Administrative Service in the remote border district of Tawang in Arunachal Pradesh. Accounts of my experiences there appeared in various articles and as a book, *Tawang–The Land of Mon,* which was published in 1982. In Tawang, in the early nineteen seventies, I saw a Himalayan paradise at first hand, almost untouched by external interventions and still governed largely by local systems for sustainable management of land, forest and water resources. These management practices extended to private as well as community property resources and my first hand knowledge of these systems has deeply influenced my approach to the problem of working out appropriate afforestation strategies for the UP Himalayas. By 1986 it was almost ten years since I left Tawang and the choice of another Himalayan district for a study leave project was inevitable. The Tehri dam guided my choice since it was but natural that afforestation should command prime attention in the wake of construction of this controversial dam which will probably be the largest earth dam of its kind in the world, when it is eventually constructed. The Planning Commission agreed to fund a small project which was sponsored by the National Centre for Human

Settlements and Environment in Bhopal. My field studies were completed between Ocotber 87 and February 88 when I had to rejoin my cadre on posting to Goa. The National Centre for Human Settlements submitted the report of the study conducted by me to the Planning Commission in 1989 but this report could not, for obvious reasons, encompass all that I had absorbed in the eventful one and half year of field work. My subsequent experiences during the next ten years of continued association with Tehri Garhwal also added to the data base of 1987. The present structure of the book thus emerged quite simply as a record of the processes of destruction of the Himalayan forests of Uttar Pradesh, followed by an evaluation of the state efforts towards reversing this trend through massive plantation programme. Special reference to the district of Tehri Garhwal marks both the sections. Gradually my data bank on people's efforts towards protection and afforestation in the same district also increased to a point where these accounts had to be grouped separately. The book is thus divided into three parts, though a vital component of the people's protection efforts, namely the chipko movement, has been omitted because the subject has already been extensively researched and written upon.

Is this study new in any way? To whom is it addressed? All throughout the long and often tedious hours while I was working on the book I often felt it would fall between two stools. It is not a purely academic work (of which there is no dearth for the UP Himalaya,) and yet it is doubtful whether the administrators, policy makers and planners to whom it is addressed will ever get down to reading it! Still, the facts had to be stated and the stories of humble people-within the forest department and in the villages-had to be narrated. I can only hope that the discerning reader will sift the chaff from the grain and utilise this book for the wealth of data that it contains, both in terms of statistics as also narrative accounts of this ongoing process of deforestation and reafforestation in the central Himalayas.

PART I

DESTRUCTION

No doubt such iron foundries have been found to be very destructive in Britain in the past where they were legally banned by the 16th Century. However, such action is deemed neither possible nor desirable here. And even if such laws were to be passed, *for whom would these forests be preserved?*

Job Becket, Deputy Collector of Kumaon, on the Kumaon Iron Works which by 1868 had been leased 400 sq. miles of virgin oak forest to serve as fuel for foundry.

CHAPTER 1

Who is Destroying the Himalayan Forests?

Who is destroying the forests of Uttarakhand — the devabhumi, valley of gods ?

At six in the morning in a teastall near Kausani, I had once witnessed a discussion on the subject conducted avidly by hillmen in ragged clothes, patched Jawahar jackets and faded caps. They were huddled on narrow wooden benches, their shoulders hunched against the early morning chill. Warming their fingers around mugs of hot tea, the men discussed the issue quietly, gravely, concluding in gentle fashion that the valley had become a battleground between gods and demons, with man's demonic nature rearing its head like a malevolent serpent threatening to swallow the valley of gods.

Who are the demons? Nobody stated it explicitly. Greed was implied with many a subtle nod. Was the demon of greed lurking within? Or were they in the grip of a malignant external force that will ultimately engulf and destroy them? Superstition and religious belief rub shoulders in Uttarakhand. Perhaps the people are governed by a latent belief that the gods will ultimately salvage their land.

Uttarakhand, or Kedarkhand of the Hindu scriptures, has always been the traditional devabhumi (abode of gods), a place of retreat and worship with a ritual and sacred status that is unique amongst the states and countries that span the Himalaya across the Indian sub-continent. In myth and religion Uttarakhand is typified by the Ganga, holy river of the Hindus, which surges out of the yawning cavernous mouth of the Gangotri glacier at Gaumukh, foaming white spray forming a high mist, like a calvacade escorting the massive torrent as it rushes down the sacred himalaya to nurture the great North Indian plain.

Nowadays we repeat the old legend of the descent of the Ganga, giving it new significance. Shiva, the Lord of the Himalaya, came forward to break the gigantic descent of Ganga from heaven. As the mighty river got entangled and dispersed in Shiva's matted locks, she began meandering her slow way down to give life and sustenance to the Indo-Gangetic plain. Thus Shiva's locks saved the earth from being washed away by the force of her torrent. We can take this legend as a symbolic explanation of the role played by the matted roots of the Himalayan forests in protecting the entire eco-system, not only of the Himalayan mountain system but also of the Gangetic plain.

But Shiva's locks have been shorn now. Like Samson, the strength of the great Himalaya, bereft of their mighty forests, is being slowly sapped away. The people who live there know it, feel it, and search for answers, looking for explanations in human greed. Being humble and honest, they do not excuse themselves either, for they too fear the demons within.

For city-based planners and writers, the task of pinpointing the destroyers seems less difficult. They can, with ease, place the entire blame for devastation of the Himalayan forests on the people themselves. Research scholars find it easy to assume that the Himalayas have reached a stage where the resource base is gravely endangered due to a combination of over-population and overstocking of cattle. This line of thought has perhaps been best expressed by D.P. Joshi, a retired Forest Officer of Uttar Pradesh.

> The Himalayan eco-system has been bled to death by those very people to whom these forests mean bread and life. They, out of lack of foresight and ignorance of the need for sustained preservation of a forest system, have almost succeeded in killing the goose that lays the golden egg.[1]

Foreign research also seems rest to primarily on the overpopulation theory.[2] International agencies finance schemes which require local population to be evacuated in order to make way for biosphere reserves and national parks in order to save endangered eco-systems.

These sweeping accusations are still understandable when they emanate from international agencies, policy planners and officers of the forest department, since the state and the forest department have had a historic confrontation with people. But such assumptions become harder to accept when they are made by writers who are otherwise sympathetic to the cause of people's struggle against state sponsored exploitation. The tendency to blame indigenous people for degradation of their immediate environment, thereby diluting the responsibility of colonial powers or state governments thus appears to emanate from a deep-rooted prejudice which often defies or ignores empirical evidence to the contrary.[3]

Another line of thought tends to be ambivalent about the role of the state as destroyer, by making a distinction between the policy and the people implementing it and subscribing to a fairly common belief that management of forestry as introduced by the British was well-intentioned and scientific, but gross and reckless exploitation of forests was triggered off by greedy politicians and bureaucrats of modern India. Forest Officers often hark back to the "good old days" of British Raj when forest management was properly implemented without being subject to democratic pressures triggered off by politicians who apparently encourage landless farmers to encroach on forest lands. Before independence, famines and abysmal health care kept life expectancy low, thus delaying the population explosion which has now increased pressure on forest resources. The great divide is thus assumed to have occurred, not in 1816 when the British annexed Kumaon and portions of Garhwal or even in 1858, but in 1947, or 1962. The post-independence period is hence seen as having brought in its wake the holocaust of urban explosion and development which has led to destruction of the forests of Uttarakhand.[4]

Conceptual confusions and lack of historical contextual evidence in approaches to the problem should have been firmly set at rest by Ramachandra Guhas pioneering work of outstanding scholarship on ecological change and peasant resistance in the Himalaya.[5] Guha documents in chronological detail the actual processes of ecological change introduced by

British silvicultural systems of forest management which led to radical deprivation of villagers and erupted in a continuous series of peasant uprisings and unrest. He highlights the existence and strength of traditional systems of management of community property resources, linking these with similar systems found amongst mountain and tribal people in India and elsewhere. The people of Uttarakhand had enjoyed age — old traditional rights over their forests which were taken away by British colonial forest policies. Infact it is these indigenous management systems by which generations of native rulers had managed to conserve the Himalayan forest resources as an inexhaustible bounty which the British received as a veritable gift in 1815.[6]

Destruction

The history of forest policies adopted by the British, as also the heritage left by their scientific silvicultural management practices, merits extensive documentation in order to trace the exact processes which led to the destruction of the enormous bounty of nature which the British inherited with the annexation of Garhwal and Kumaon in 1815. The first part of this book investigates these processes of destruction.

The three chapters that follow present statistics culled from the earliest working plans and gazetteers of Garhwal to show how the forests, the pastoral economy and the entire eco-system of Garhwal were destroyed by the policies of the East India Company and the Imperial Government. While presenting the data gathered from these sources an attempt has been made to also show how so-called scientific forestry was based largely upon conjecture and economic interests of the age, rather than on any scientific hypothesis.

The transition after Independence was smooth if not imperceptible. This take over was further consolidated by the National Forest Policy Settlement of 1952. The latter paid obeisance to the functions of forests for environment protection and for providing subsistence to the local people who depend upon them, but in effect, dedicated timber solely to the needs of 'Industry and Development.' The policies and systems introduced by the British thus continued, and the processes of destruction inherent in the systems of forest management were given further scope for progression even though lip service was paid to the need for eco-protection. Hence the independence period is logically included in the first part. Since the statistical base of the department also remained strong, the actual trends of forest policy emerge easily when forest statistics are studied in detail.

Documentation of the destruction of the forests of U.P. hills after independence has relied largely upon the Forest Statistics of U.P. 1983, while a longitudinal profile of the reserve forests of Tehri Garhwal has been based on an in depth comparison of statistics found in the working plans of P.N. Khosla 1940 and A.M. Rai 1981. In a way, the latter exercise has revealed the results of the silvicultural management system of the forest department over a period of four decades.

In addition to the above, considerable amount of data was personally collected from the office of the Tehri Forest Division during the writer's one year sojourn in Tehri. This part of the study gives an update till 1987 and incidentally, also presents some disturbing facts regarding forest management in the post 1981 period after enforcement of the ban on felling of green trees above an altitude of 1000 meters.

Restoration

The people's side of the story had always simmered beneath the surface. People's protests had started occurring in British Garhwal as early as 1893. This protest was most ably expressed by G.B. Pant in his historic treatise on the forest problem of Kumaon in 1922.[7] In the decade of the seventies the Chipko movement swept through Garhwal and succeeded in dramatizing the plight of the local people who have had to struggle for existence in an environment ruthlessly exploited by commercial interests. This resulted in the famous ban imposed on cutting of green trees in the Himalayas above an altitude of 1000 meters.[8] Restoration now followed destruction, but without the regularity and certainty of natures cyclical rhythm.

The second part of this book has accordingly focused on reafforestation policies, survival rates of government plantations and the various options available for greening the Himalayas. The government plantations have been reviewed in terms of the performance of plantation divisions working in the catchment of the Bhagirathi river in Tehri and Uttarkashi falling under the Conservator, Bhagirathi circle.

An attempt has been made to strike new ground by viewing the process of regeneration separately in terms of peoples efforts and government programmes. Documentation of people's efforts for protection of village forests was more challenging, but also a rewarding task. The last part of the book is devoted to an analysis of how mountain communities traditionally managed their community property resources across the Himalayas from Uttar Pradesh and Nepal to Arunachal Pradesh. The history and record of traditional property rights in the U.P. forests is also sought to be brought into proper perspective. Thereafter, narratives as well as statistical accounts of peoples protected forests, identified by the author over a ten year association with the district, are presented to show how numerous villages in Tehri Garhwal still have indigenous management systems which cover scattered remains of reserve and revenue forests being voluntarily protected and managed by the communities themselves.

The second and third parts of this book taken together focus on the processes of restoration—the sarkari (government) way and the peoples way, while assessing the success and the failures of these two parallel movements *viz.* the forest departments efforts to plant up 'new' forests and the villagers isolated but widespread and determined ventures to protect and thereby restore 'their' old, often decimated, forests. Readers will have to judge for themselves which of the two strategies would be more effective in saving the forests of Uttarakhand.

The Wheel of Time

The fact that the wheel of state policy invariably lurches towards a full circle was brought home in a telling manner one winters morning in January 86 when the author accompanied DFO Civil Soyam to inspect a walnut plantation which he had laid out with love and care on a plot just adjoining the famous Chandrabadani temple of Tehri Garhwal. The plantation site was marked by spindly oak scrubs. No villager accompanied us. The single chowkidar prayed humbly for an increase in the princely sum of Rs.100 being paid to him as a monthly wage. (The philosophy of the National Wasteland Development Board has been that part-time care should suffice, since they fondly believe that thc village community will take care of plantations which government is good enough to set up for them).

In Chandrabadani, nobody cared to recall events which had taken place, almost 82 years ago, at the very spot where the forest department was now painstakingly laying out these walnut plantations. In those days the Chandrabadani temple had been sited in the midst of thick oak forests widely regarded as sacred, while being valued also for leaf fodder, compost and the protection they afforded to natural springs whose waters provided irrigation and the necessities of the village itself. These sacred groves characterised almost all temple tops besides the main sacred tirthas of Kunjapuri, Surkanda Devi, Ghandialdhar. The existence of these groves is not confined to Garhwal and Kumaon, since similar sacred groves have been widely documented in Karnataka, Maharashtra and Orissa and the tradition itself perhaps reaches back well over four thousand years. After 1899, when demarcaion and notification for reserving prime forest land started in Tehri state, on the pattern of forest administration in British Garhwal, the forest department attempted to demarcate and expropriate the oak forests of the Chandrabadani temple. The forest team was led by the Conservator. On 27 December 1906, a historic confrontation occurred in which the entire Jakhanidhar area gathered, rising up in great strength, armed with arrows and lathis to drive out the surveyors. The Conservator was attacked and branded with a hot iron. The Raja appealed to the British for help. Like the better known Rawain episode, this peoples resistance also came to nought.[9]

By expropriating the oak forests of Chandrabadani the state had not merely destroyed a forest but had also shattered a social system, based on magico-religious perceptions, by which the entire area co-operated to maintain and manage it's micro-environment and community property resources. Over a period of seven decades, alienation from their sacred and invaluable forests gradually set in. An area which had once belonged to the Jakhanidhar region as a whole and had been venerated as a sacred space by all villagers, now turned into a no man's land which everyone exploited. Nobody cares to protect it now. The same villagers who once arose in desperation, branding the Conservator with a hot iron and taking to arms in defence of the Chandrabadani forests have, become utterly indifferent to their fate over a period of eight decades. It is not surprising that only oak scrub remains and the villagers take no interest in the government efforts to raise plantations in the same area which the state had once confiscated from them.

It was a sad and instructive experience to see how the wheel had indeed turned full circle in less than 85 years. The state which once forcibly denuded and deforested these temple groves, had now come back in the role of saviour. Eight decades ago, villagers were regarded as culprits. Today also villagers are, in a sense, viewed as culprits. The state comes in paternal guise to elicit 'community participation' for social forestry programmes 'meant for the people.' Urbanised students like this writer come to study and monitor their progress.

The expropriated are now urged by the expropriator to come forth and take charge of their ravaged lands — to unite, co-operate, provide social fencing. Villagers views, needs and rights were disregarded at the time of expropriation. Afforestation programmes too tend to pass them by.

NOTES

1. Joshi, D.P., "Forest Policy for the Regeneration of the Himalayan Environment" in *Environmental Regeneration in the Himalaya Concepts and Strategies:* (Ed) J.S. Singh.

2. Ives, J.D. 1985. "Mountain Malaise", in *Integrated Mountain Development,* (ed) T.V. Singh and J. Kaur.
   Reiger, H.S. 1981. "Man Versus Mountain", in *The Himalayas : Aspects of Change.*
   Pande, D.C. "Forest Management", in *Rural Development in the Himalayas.*
3. Thomas Weber. *Hugging the Trees;* Penguin; 1989 Ch.1; p.19. The writer quotes a traveller Bishop R. Heber, writing in 1824-25, to support the thesis that "great devastations are generally made in these woods partly by the increase in population and partly by traveller and cattles and goats." He further suggests that the *local population* under colonial rule removed the best quality trees for their own usage followed by *villagers* collecting fire wood and grazing live stock. There is no mention of the wasteful and exploitative policies of the British Raj and their system of silvicultural management. Apparently the writer owes these views to R.P. Tucker, "The Forests of the Western Himalaya" Journal of Forest History, 1982; 26(3).
4. *Ibid;* p.23; Weber states that the fate of Uttarakhand was finally sealed in October 1962. According to him the feared invasion came from the Indian South not from the Chinese North.
5. Ramachandra Guha. *The Unquiet Woods; Ecological Change and Peasant Resistance in the Himalaya;* Oxford University Press; Delhi; 1989. (hereafter *The Unquiet Woods).*
   Amar Farooqui. *Colonial Forest Policy in Uttarakhand 1890-1928*; Kitab Publishing House, New Delhi, 1997, also combines the social and environmental history of the Kumaon region with a systematic analysis of the impact of the colonial regime on the traditional symbiotic relationship forests had with cultivation and the lives of common people in pre-capitalist societies.
6. E.T. Atkinson. *The Himalayan Gazetteer; Vol I Part II, Vol III Part I.* Reprint by Cosmo Publications, 24B, Ansari Road, New Delhi 1980. (hereafter *Gazetteer*)
7. Pant, G.B. *The Forest Problem in Kumaon;* with commentary by Dr. Ajay S. Rawat; Gyanodaya Prakashan, Nainital, U.P.
8. There is extensive literature on the Chipko Movement. The most authoratitive work which also places the movement in the proper perspective of peasant struggle all over the world is that of Ramachandra Guha. Guha also dwells in detail on the silvicultural management practices which directly led to the expropriation of the people and destruction of the forest wealth of Uttarakhand.
9. Guha, *op. cit.*

CHAPTER 2

British Forest Policy in the Nineteenth Century: The Formative Years

What were the main features of British forest policy which evolved over a period of time to destroy the ecology and economy of Kumaun and British Garhwal? Two phases can be traced: 1818-1859 when the forests were ruthlessly, if not mercilessly felled by the East India Company and the period of 'conservancy and scientific management' which set in with the replacement of the company by the rule of the Crown in 1858, followed by establishment of the Imperial Forest Department of the North West Provinces in 1868.[1]

Towards 1868, or the last quarter of the nineteenth century, after destruction had reached its peak, so to speak, during 50 years of British Rule, under the East India Company, the Imperial Forest Department came into being in the U.P. Hills to scientifically manage and curtail this process. A series of pioneering surveys of the forest wealth that remained in the coniferous belt were undertaken in the middle and upper regions, after which commercial extraction went into full swing.[2] It is significant that extraction continued without any working plans for another 25 to 50 years. The first working plan of British Garhwal (for the middle and upper regions) was drawn up by A.E. Osmaston only in 1919, almost half a century after the establishment of the Forest Department. It would be a mistake to assume that history of forest exploitation by the British commenced with Osmaston's first working plan of Norh Garhwal, 1921-22. The truth is that, even after the setting up of the Imperial Forest Department in 1868, many forest areas were extensively clear felled and overworked for over half a century of British rule without any clear rotational system. Fortunately, skeletal Working Plans were drawn up in the 1880's for the sub-Himalayan tracts which were formed into the Ganges and Garhwal Forest Divisions. From these working plans we can glean a great deal of information regarding forest policies in the sub-Himalayan sal tracts during the days of Company Raj and just after, till the time of establishment of the Forest Department in 1868. These plans bear witness to the fact that the sub-Himalayan tracts had been subjected to forest operations under 65 years of East India Company rule without any ground rules, conservancy or care. It is this half century and odd of outright plunder which shook the ancient forests like a cruel storm, destroying it from its ancient roots and the British records themselves speak of the mindless destruction and waste which occurred.

The history of forest operations in British Garhwal actually began around 1840 with the onset of the need for commercial exploitation in response to railway construction activities in the vicinity of Haridwar and Nazibabad. Later, a further impetus was given by construction of the Upper Ganga Canal, along with development of big timber markets in Meerut, Saharanpur and Haridwar. The main pressure on forests, though, was occasioned by railway construction works which reached a peak after the first major Indian Revolt of 1857 (more commonly known in British history as the Indian Mutiny). The heavy demand for sleepers was met in the early years by rich, invaluable sal forests of the Doon Valley. In the Gazetteers of the Himalayas, Atkinson clearly states that the latter were soon destroyed and that it is

> only after denudation of the thick foothill jungles that the British authorities realized the value of immensely rich deodar forests located in the Himalayan belt of Jaunsar Bawar, north of Dehra Dun near Chakrata.

Subsequent years were, therefore, marked by a consistent trend towards notification and expropriation of deodar forests in Jaunsar Bawar[3]. Surveys and commercial exploitation of chir pine forests was also started in British Garhwal. By 1868 the railway department had already inspected and approved cutting down of chir pine for railway sleepers as far north as Mandakini valley, as also further north near Pandukeshwar.

The logs were floated down the Mandakini and Alakananda rivers to Rishikesh. Earlier still, around 1850, Mr.Wilson had obtained his contract of dubious fame for a similar purpose from the Raja of Tehri. This lease led to heavy destruction of unbelievably thick and untouched Deodar jungles around Harsil along the Upper Bhagirathi river, main source of the mighty Ganga.[4]

In 1853, the foothill forests of Garhwal and Kumaon fell under the charge of Capt. Reid and one mysterious Mr.Finn and were worked by them till 1858. These two names have become synonymous with complete devastation in all subsequent reports of the 19th century.[5] Areas in the sub-Himalayan tracts worked by Capt. Reid for the gun carriage agency in particular, included whole tracts of the Taimuriya (Palain) basin in the Patli Dun, and were overworked and thinned out by him. In the same period, according to Atkinson, forests of the Ramganga valley (lower) Patli Dun (South) and the Sona were *'felled even to desolation'* and many parts

> *have been permanently injured,* the land where once fine sal forests stood is now too denuded by exposure to admit of reproduction.[6]

Capt. Reid had even established a saw mill in the forests of Boxar and started logging operations on such a large scale that vast, in fact enormous,quantities of dead timber remained lying in the forests long after his departure.[7] Revenues realized from the Patli Dun forests must have been considerable, but Atkison's Gazetteer is silent on this point, perhaps because systematic record keeping had not yet started.

In the same period Kotri Dun was drastically overworked in all its tracts and only a few good forests remained in the remote and inaccessible regions. These remained under the lessee system till 1868 *i.e.,* till the establishment of the Forest Department.

In the Kumaon foothills (comprising large parts of the Ganges Forest Division which was formed in 1888) heavy felling took place along the Rawasan Nadi and, if we are to go by the compartment descriptions in Hearle's Working Plan of 1881-82, little valuable timber was to be found in the area when the working plan was drawn up. Several parts of Rawasan and Sigadi blocks had nil, rare or indifferent reproduction due to being overworked or subjected to uncontrolled fires.[8] Atkinson states that during the entire period of Company Rule:

> the forests of the present Kumaon Forests division were denuded of good trees in all easily accessible places. Between 1855-57, demand for railway sleepers induced heavy speculation and reckless cutting of old trees far in excess of what they could export, so that for some years after regular forest operations commenced, the department was chiefly cutting up and bringing to the depot the dead timber left.[9]

About forests in valley of the Ramganga, the South Patli Dun and Sonali, Major Pearson the first Conservator of North West Provinces (which included UP hills) wrote:

> In one word they may all be said to have been felled in even to desolation by Capt. Reid as well as by Mr. Finn, and the native contractors who went before him, *but perhaps even this does not give an idea of the waste that has occurred and the mischief that has been committed.* Thousands of trees were felled which were never removed nor was their removal possible; and a large revenue has been realized during the last few years by allowing passes to the people of the lower country to cut up and remove the dead timber on the payment of a royalty.[10]

One of the worst side effects of these operations was the commencement of huge forest fires which were inevitably ignited by the massive quantities of refuse and perfectly serviceable timber which was left to rot in the forest. Major Pearson, in his report of 1868 and Hearle and Dansey in their working plans of 1888 and 1891 continued to bemoan the effects of repeated fires on the long term health and prospects of the forest. To quote Dansey:

> Fires deteriorate the soil by consuming the leaf mould which goes to form its most nutritious part. Again it hardens the soil, so that air and water can only introduce themselves with difficulty, thus rendering the formation of new soils a slow process, as also the further decomposition of the elements of the soil itself. Indirectly, fires injure the soil by hurting the vegetation that covers it, by tending to its denudation, to its being washed away or baked by the sun.[11]

The position in Dehradun was perhaps the worst. The earlier native system prevailing under the Garhwal Raja had ensured that purchasers from the plains only paid toll for produce which they bought and took out from the hills, while the hill people themselves remained masters in their own terrain.[12] Exploitation of the forest produce had followed traditional boundaries, where village control over common lands existed and, therefore, acted as a check to over-exploitation, particularly by merchants and banias or contractors, who never had any stake in preservating the natural wealth of any region. In 1816, when the British took over, the Chauki or forest toll gate of Dehra Dun was collecting, as per the policy of the native state, Rs 16,000 annually on outgoing forest produce- no mean sum in those days.[13]

While this chauki or toll gate was abolished, the lessee system introduced by the British Raj in its place proved to be disastrous on all accounts. The forests of the Doon were apparently the most lucrative. The British let out the forests to contractors in a coupe or 'Kathbans' mahals (as they were called) at an annual lease of Rs.4,000. It was to be expected that, having 'bought out' the rich forest of the Doon, the contractors and leaseholders from the "lower country" proceeded to take out what they could, making the best of a pretty good bargain. As Atkinson states

> Everyone continued to hack and hue away as he pleased. Fine trees from 100-200 years old still abounded in the district. All these fell before the axe and probably the rest would have gone with them had the roads been better.[14]

The British overturned the indigenous system of exploitation of minor forest produce by local resident communities, since they always had a particular inclination to deal with the locals of this country through middlemen, be they zamindars or contractors. Indigenous rights were extinguished and outsiders brought in to deal with the forest. The classic era of the middleman had started.

Meanwhile, the oldest department of the British Raj — the Land Revenue Department — had also been following policies which encouraged expansion of cultivation to gain mounting land revenue at the expense of destruction of forests. Massive clearance of dense Bhabbar tracts of Kumaon foothills was a notable example of this process.

After this period of devastation by private contractors, whether by lessees or Mr. Reid and Mr. Finn, efforts at conservancy set in around 1858 after the British Government took over from the East India Company. North Patli Dun, Kotri Dun and other forests which were later merged into the Ganges Forest Division were brought under care and management of Sir Henry Ramsay, Commissioner of Kumaon and ex-officio Conservator. For ten years Ramsay did what he could for these forests, even though it meant that they had to be closed completely to the hapless villagers who were made to shift out along with their cattle stations.[15] It is significant that closures to man and cattle had to follow upon reckless commercial exploitation. It is the latter and not the former which destroyed the sub-Himalayan forest, the costs of which had to be borne by the villagers and the pastoral economy of the region, and later, much later, by the entire sub-continent.

On the eve of the establishment of the Forest Department in 1868, future plans for working and managment of the forests were summed up by Major Pearson in a status report.

> After the inspection I have made I believe I may safely commit myself to say that Col. Ramsay's estimate of one lakh cubic feet per annum (annual yield) from the forests of both Garhwal and Kumaon divisions may confidently be reckoned as a safe one.[16]

In addition to the above there would be removals of partially unsound trees and from thinnings.

Along with heavy working of forests, followed by closure and rest, Major Pearson adopted a policy was of opening up of roads in every case in addition to the network which had been established by Col. Ramsay. This greatly facilitated removal of dead wood, bamboo and other minor forest produce by the people of the 'lower country' for sale in the big markets of Roorkee and Nazibabad. As far as the forest department was concerned, there were only two potential enemies or rivals for usage of forest products — the revenue department and villagers. With the former an uneasy truce was declared, by handing over to the jurisdiction of the revenue department those residual forests (later designated, ironically perhaps, as Class I Forests) "where" (in Waltons words) "hardly any forests remain".[17] The revenue department was to grant pattas or leases for further expansion of cultivation only in these lands. As for the villagers themselves, systematic and planned expropriation was the only answer. As always, somebody had to pay the price for the prosperity which suddenly flooded the British colonial government and who else could pay it, except the hapless inhabitants of the Uttar Pradesh Himalayas?

Revenue receipts and net profits

What were the takings on the British side?

In the early years it is the lessees who took the cream of the earnings, facilitated no doubt by the confused and ad-hoc method of administering these mahals which were variously supervised by the Commissioner Kumaon, the Collector of Bijnore, and Collector of Saharanpur (the latter of whom worked through the Superintendent of the Doon). It was the Superintendent of Doon who first woke up to the heavy losses being incurred by the British (then the Company) Raj through the lessee system. Rents had languished while the lessees profits must have increased considerably since, when publicly auctioned, the lease was fetching Rs.35,000 per annum. In the early eighteen fiftees, the Superintendent of Doon realised that the lessee, while paying them about a lac of rupees per annum, was still netting a profit of over Rs.80,000, and so he quietly terminated the arrangement thus appropriating the extra profit as well.[18] By 1855, when the Dehra Dun forests first came to be managed directly, therefore, the profits would have been over one and a half lac of rupees per annum. This was indeed a stupendous sum. Unlike the Garhwal and even the Gurkha Rajas whose needs were modest and who were therefore content with a smaller but steadier income, the rapaciousness of the Company Raj (which had to satisfy needs, not only in India, but also back home) could not be easily contained. Revenues thus (as reported by Atkinson) "rose enormously" after 1855, when they were in any case over Rs.1.50 lacs per annum. At a conservative estimate, the figure might have doubled to Rs. 3 lacs per annum. But since there was no conservancy worth the name, the heyday lasted only till 1867-68, when the revenue fell back to Rs.23,332 per annum.[19]

Available figures for the Bhabbar and foothills tracts culled from Atkinsons Gazetteer and Working Plans are given below[20]:

Forest Revenue 1858 to 1908

(rupees in lacs)

Forest Division	*Period*	*No. of Years*	*Gross Revenue*	*Net Surplus*
Garhwal	1858-59 to 1867-68	10	18.18	10.02
	1868-69 to 1879-80	12	17.89	8.79
	1889 to 1894	6	11.76	8.10
	1898-99 to 1907-08	10	7.50	4.30
Kumaon	1859-60 to 1867-68	9	15.00	6.68
	1868-69 to 1879-80	12	25.84	15.22
Dehra Dun	1865-66 to 1879-80	15	9.31	2.64

The total net surplus of the incomplete data given above amounts to over Rs.58.0 lacs without counting Capt. Reid's period from 1854-58 (which should have netted Rs.3.0 lacs per annum) and the period of reckless felling in Dehra Dun after 1854 when revenue "rose enormously". Since the revenue under the lessee in 1854 was Rs.1.80 lacs, one can presume that the annual average surplus of Dehra Dun could have been roughly Rs.2.5 lacs till 1865. This reckoning by itself gives a total of over one crore rupees. Once the missing figures are taken into account, it may not be an exaggeration to suggest that in 50 years of British rule from 1855 to about 1908, the sal forest tract of the sub-Himalayas in Uttar Pradesh had alone yielded well over one and a half crore rupees to the Imperial exchequer. This gives a reckoning of about an average of Rs.3.0 lacs per annum - not an inconsiderable sum at a time when the daily wages fluctuated between one and one and a half annas, rising to two and a half annas in 1896.[21]

The opening up of roads, the extinction of the hill people's age old rights over forest produce, grant of passes to contractors and middle-men from the plains for commercial extraction and the vast expansion of urban markets — in these policies there lay the seeds of forest management practices which were to govern India for the next 150 years.

This three-pronged strategy of forest operations for commercial profit started by British rule was now to prevail in the Central Himalayas .

How scientific and effective were these management practices in achieving atleast their declared objectives? The next chapter tries to answer this question.

NOTES

1. Ramachandra Guha. *The Unquiet Woods* Oxford University Press 1989 p. 35. Guha divides the history into 4 phases: 1850-1878 when the state concentrated on extraction in the sub-Himalayan region. 1878-1893 when all waste land was notified and taken over by government as protected forests vide the notification of 17th October 1893.
   1893-1903.
   1911 onwards. The data regarding resin production and fire protection pertains to the post 1907 period.
   Ajay Singh Rawat *Garhwal Himalayas:A Historical Survey (1815-1947)* Eastern Book Linkers Delhi 1983.
   Rawat starts the history of forest administration from 1868 and takes the first phase from 1868 to 1893.
2. T.W. Webber. *The Forests of Upper India* (London 1902).
3. E.T. Atkinson. *The Himalayan Gazetteer,* Reprint, Cosmo Publications, New Delhi 1980, Vol. I, Part II, pp. 833-834, pp. 869-872.
4. G. F. Pearson. *Report on the Bhagirathee Valley Forests,* Selections from the Records of the Government of the North West Provinces 2nd series,Vol. 2 (Allahabad 1869).
5. Atkinsons. *Gazetteer,* and Major Pearsons reports contain several graphic accounts of the damage wrought during this period. Also see: F. Beadon Bryant, *Working Plan of Garhwal Forest division 1896* . E. P. Dansey *Working Plans for the Garhwal Forest division 1879-80.*
6. Atkinsons. *Gazetteer* Vol. I, Part II p. 861.
7. *Ibid* p. 858. Also see F. Beadon Bryants *Working Plan.*
8. N. C. Hearle. *Working Plan for the Ganges Forest Division, 1896* pp.78-85, also Atkinson,Vol. I, Part II p. 862.
9. Atkinson. *op cit* p. 852.
10. G.F.Pearson. "Sub-Himalayan Forests of Kumaon and Garhwal" in

Selections from records of the Government of the North West Provinces, 2nd series, Vol. 2, Allahabad 1869.

11 E. P. Dansey *Working Plan.*

12. Atkinson. *Gazetteer op cit* p. 845.

13. *Ibid;* pp. 863-64.

14. *Ibid;* p. 864. This quote is from memoirs of Mr. Williams, Superintendent of the Doon.

15. *Ibid;* pp. 858-59.

16. Pearson *op. cit.,* p. 132.

17 H. G. Walton *British Garhwal: a Gazetteer;* Allahabad; 1910; Reprint Nataraj Publishers, Dehra Dun; 1989; p. 17; Walton states that Class I comprise forests which had practically disappeared and were not worth saving and were made over to the revenue department for management. In almost in the same breath he stated that the district forests were to be administered solely for the benefit of the people.

18. Atkinson *Gazetteer op cit* pp. 863-64.

19. *Ibid.*

20. *Ibid;* pp. 916-117. Also see Beadon Bryant p. 20, 28.

21. Beadon Bryant p. 31.

CHAPTER 3

Destruction of the Eco-system

One of the most personal and vivid expressions of the philosophy and avowed aims of the Imperial Forest Department, at the time of its establishment, can be found in Major Pearson's report of 1868 on the sub-Himalayan forests. This bears quoting at some length:

> In those forests which were partially but not too heavily worked in past years. The growth of the young trees and saplings in almost every case is most satisfactory.. when one looks at thriving young trees of every size.. all full of the highest promise, the mind is filled with good hope that future ages will see these valleys and plateaus covered with forests far greater in value to those that went before them, in as much as they will benefit by regular protection and conservancy which their predecessor did not enjoy.[1]

Further extracts from Major Pearson's report shed light on the principle of so called scientific management which were adopted:

> I have now been in the habit of watching sal forests for about ten years and the inspection of these Doons very much confirms the opinion to which my mind has for some years been tending — *viz.* that both for the free generation of the seed and the effectual reproduction of the forest, as well as for the welfare of the trees, and their progress afterwards, *sal requires a considerable amount of sun and light* (italics in original) and that a sal forest will bear, and indeed repay (if it does not absolutely require), much more liberal felling than almost any other description of forest in India.

Hence the Conservator laid down the prescription that each sal forest block "should be worked in succession to "the full extent it will bear with safety". He felt

> no hesitation in saying that if half the trees were cut down in the untouched portion of the Palein forests as well as in the Mondhal and Nindhore valleys, the remainder would benefit beyond all calclations by operation.[2]

In brief, 'protection and conservancy' in British forest department parlance meant silviculture operations which involved clear felling of pristine forests and conversion to uniform age classes thereafter in order to further monoculture of sal or pine (as the case may be). Naturally, clear felling had to be followed by complete exclusion of cattle and human interference. Fire protection was to be achieved by regulated and systematic use of fire as a tool of forest management.[3] Miscellaneous forest species were to be reduced, if not eradicated, in order to make way for the advance growth of commercial species *which were not always the dominant species.* In the northern belt, the broadleaf species formed the climax forest but had to give way before the commercially valuable chir pine.

While making these prescriptions the Working Plans were confident that over a period of scientific working of two or three such plans, even better forests than the ones existing would be left for posterity.

Were these hopes fulfilled? Was their main aim of advancement of monoculture of sal and conifers at least achieved, even though it had to be at the cost of destruction of so many other species and further deprivation of the local people?

The Patli Dun forests of the sub-Himalayan tract may be taken as a test case since Major Pearson himself testified, in 1868, that they still retained considerable tracts of excellent sal forests after the depradations of Company Raj. It may not be out of place to also mention his certificate that "*there is no State forest in India to be compared to this one for extent, compactness and perfect control*".[4] Hence if the forests did not radically improve after three decades of scientific management, the blame cannot be laid (as it almost always is laid) at the doors of the local population. Let us, therefore, turn to E.P. Dansey's Working Plan of Garhwal Forest Division drawn up in 1898 where, contrary to Major Pearson's glowing predictions, a rather grim picture of the same area emerges. After thirty years of conservancy operations, Dansey bemoans the fact that

> over much of the working circle a soil which, from the dimensions of the standing trees we see, must have been very rich originally,

> has since then undergone deterioration through *denudation which has been brought about by excessive fellings and long exposure to the sun.*

That poor soils were very extensively found is borne out by the fact that out of an area of 79 sq. miles, only 36 sq. miles could be said to be very suitable for good forest vegetation.[5]

An extremely important observation was made by Dansey with regard to silvicultural operations on sal. In his division he found that on colder aspects it was common to see considerable damage caused to young sal poles by severe frost. He concluded that

> this should teach us not to create blanks or *even to open the forests more than would be needful for the admission or actual quantitites of light required by the circumstances of the case.*[6]

Dansey, however, could not lay down the law for the Imperial Forest Department which continued to follow prescriptions to open up sal forests to light by clear felling of first class trees as also by eradication of miscellaneous or 'inferior' species of trees, shrubs and climbers.

The course of events proved Dansey to be correct, rather than Major Pearson. In 1898 only small blocks of sal were noticed by Dansey as being severely affected by frost. By the 1930's, Lansdowne Forest Division had to have special prescriptions for nursing and care of whole blocks of sal forest which were separately designated as "Frosted Sal Working Circle". This problem (which never came to notice before the advent of British Rule) was later to become an endemic and perennial problem of sal forests, not only in the Uttar Pradesh sub-Himalayas, but also in the extensive and rich sal forest tracts of Madhya Pradesh.[7]

Worse was yet to come in the central Himalayan oak forest belts. In order to further the advancement of chir, systematic girdling and destruction of invaluable broadleaf species of North Garhwal was advocated by A.E.Osmaston in his Working Plan of 1921-22. Appendix 17, on Girdling and Cleaning rules of the Working Plan laid down detailed instructions which were to be meticulously followed by foresters. It is interesting to note that Appendix 17 enjoins the foresters to leave a few chains constituting a narrow strip of broadleaf forest alongside each rauli, or mountain stream while eradicating the remaining trees, of course, at minimum cost to the exchequer. Thus it enjoins the department to encourage villagers to girdle oak trees and lop them as far as possible to extinction in order to further advancement of chir.[8] The same author, however, had in the very first page of the Working Plan, stated that

> it is the Himalayan broadleaf forests which protect the myriads of mountain streams which go to maintain the village 'sera' and the water system of the hills which in turn goes to feed the Gangetic canal.[9]

Ignorance can excuse many a crime, but British foresters were well aware of the role played by broadleaf forests in the Himalayan eco-system — and yet they advocated the reckless destruction of broad leaf species for short term financial gains only, leaving the Himalayas to their inevitable fate. Since villagers depended on these broadleaved forests for their livelihood and sustenance, with their rapid disappearance through systematic girdling, even the remnants of oak forests along mountain streams came to be eventually lopped to extinction. This ultimately led to the situation that prevails today marked by widespread water scarcity and drought—a scarcity which has, in turn, completely destroyed the agricultural system of the hills.

A well knit cyclic relationship can be established—broadleaf forests maintained the water system of the Himalayas, which in turn supported agriculture and good harvests, to which leaf compost and cattle manure contributed significantly. Maintenance of large herds of cattle was made possible only by the existence of extensive broadleaf forests where the system of trance humance could be practiced. Cattle were also supported by good yields of fodder from paddy and wheat straw, thus establishing a tight, interlinked circle in which the Himalayan eco-system flourished.[10]

Through reckless destruction of broadleaf forests undertaken by Government fiat, the British Government not only destroyed the eco-system and local economy, it also failed in its avowed objective of advancing the monoculture of chir pine. This failure of their system can be established by citing one particular phenomenon, that of mass felling of chir pine by storms of chir pine in which sometimes, a single storm is seen to destroy over a lakh of pine trees in a contiguous belt in areas which have been brought under scientific management by British Colonial Rule. Such mass destruction during storms is often reported in Garhwal and has been personally witnessed by this writer in 1982-83 in Mori-Naitwad area of Purola sub-division in Dehra Dun district.

While various factors such as overtapping for resin, and excessive damage by fire, have been cited as causes for this phenomenon of mass uprooting, two features which invariably characterise the affected areas deserve mention. Firstly, mass uprooting generally takes place where density of chir pine is fairly good (over 0.5) and the age class is more or less uniform. Secondly these areas are curiously marked by a singular absence of other allied as well as broadleaf species (which is to be expected in areas that have been scientifically worked for monoculture of chir). The clue to the reasons for this phenomenon of mass felling of chir pine has been found by Richard St. Barbe Baker. In his autobiography "My Life My Trees", he describes estates in Scotland which he visited in the aftermath of a storm that had blown for eight hours, uprooting eleven million pine trees. In the course of his inspection of the tree roots, St. Barbe Baker discovered that most of the trees felled were of a uniform age class (one of the marked features of the conversion to uniform system). They had simultaneously reached a stage where root competition for nutrition was at its height. Since hair roots of pines are charged with an acid sheath, St. Barbe Baker found that in monocultured tracts the pines rootlets (all from trees of the same age class) tended to form an acid pan which weakened the root system, thus facilitating mass uprooting by storms. On the other hand, he found that, in the same region, natural Caledonian forests which had a mixture of pine and silver birch, hardly suffered any casualties in the same storm. He concluded, therefore, that in a mixed forest a process of biological interaction takes place by which leaf fall of broadleaved trees forms a fungus that feeds and supplements tiny roots of pine trees which are accordingly better equipped to withstand stress.[11]

It is not surprising, therefore, that British rule could leave no significant heritage behind in the Himalayas. In 1868 Major Pearson had measured a tree in the Ramganga Valley 21' 6" in girth. But in the 1930's the Lansdowne Division was just earning its way through with the sale of young poles. It is a strange coincidence that satellite imagery of 1980's has revealed the worst erosion of the Ramganga and the Nayar catchment in the sub-Himalayan tract where forestry operations of the Company Raj and the Imperial Forest Department were first set in motion.[12]

One of the saddest tales of destruction was that of miscellaneous species. Destruction of myraids of forest species was undertaken deliberately by the Imperial Forest Department in order to further the monoculture of sal in the sub-Himalayas and coniferous species in the northern belts even after working plans were drawn up. Working Plans of the nineteenth century specifically detail hundreds of acres in which myriads of creepers and climbers were to be cut and cleared in order to further the advance growth of sal. The great variety of miscellaneous forest produce of the Himalayan districts had provided the base for a flourishing and active trade in forest products with the plains people in the nineteenth century. The list of products, staggering in its astonishing range and varieties, can be found with full details in the now rare work produced by Atkinson 'Notes on the Economic Products of the North West Provinces'. An interesting and detailed abstract can be found in Atkinson's Himalayan Gazetteer.[13] About a dozen varieties of fiber, rope making and paper making material, dyes, tanning, lacquer finish, innumerable medicinal herbs, narcotics and drugs, all types of berries and fruits, not to mention honey and falcons, are listed as being obtained from the forests and sold at high rates to traders from the plains. Even in 1915-17, eight seers of khabasi or Bhutia Badam (hazelnuts) fetched one rupee while falcons were priced at ten to fifteen rupees per bird. Osmaston notes that by 1915-16, Khair trees were over exploited for making katha. Hand made paper was a major export both to Tibet as well as to the plains. Income from sale of bamboos, ringal (cane) and bhabbar grass was very high, and generated cash surpluses, not only due to the value of the raw product, but also from allied manufacturing industries which produced the famous ropes from bhabbar grass, bamboo and cane articles from ringal etc.[14]

The raw material was generally gathered by local people during winter when the great pastoral migration used to take place to the foothills. While there was no tax on local produce and local consumption, export to the plains was taxed by the Garhwal Raja.

These forest products had sustained a large number of cottage industries which generated a sizeable amount of cash, giving the local economy a significant degree of buoyancy. Since these products regenerated annually, sustainable yields were assured, provided they were properly gathered and not over exploited. The produce was also of high value and low volume, being easy to gather and transport along hill tracks, while fetching a very high return for the labour invested. Wastage in extraction and transportation was also minimal.

The history of revenue received under the head of minor forest produce after introduction of scientific forestry tells a brief and sorry story. From it one can gauge the process of eradication of a major source of income for local people, as also of destruction of the micro-environment of the sal tracts.

Revenue receipts from sale of timber in Atkinsons period amounted to an average of little less than Rs. 3.0 lacs per annum, out of which about Rs.24,000 per annum was still contributed by the astonishing variety of forest products (excluding bamboo) listed by Atkinson.[15] Extensive details of these items can also be found in N.C. Hearle's Working Plan of the Ganges Forest Division (1888), but by that time the Ganges Forest Division could realise only Rs. 2,400 under the head of minor forest products. The larger Garhwal division (which had been partially saved from the depredations of Messrs Reid and Finn and the lessee system) was still yielding a revenue of Rs. 9,000 per annum around 1896 under this head.[16] By the time the Gazetteer of Garhwal was written in 1910, this revenue had sunk to a low of Rs.4,200.[17] Thirty years later this source of revenue had completely vanished. The Working Plan of Lansdowne Forest Division states that the revenue from the head of minor forest product was sold off along with that of the adjacent (old) Garhwal Division forests and in 1936-37 amounted to the princely sum of Rs.270.[18]

Bamboo, cane and bhabhar grass were still being sold to contractors in the Lansdowne Forest Division in 1936, fetching a modest annual revenue of Rs.1,500. This was a terrible decline from 1864-1868 when receipts from bamboo sales alone had amounted to Rs. 1,60,000 in the Garhwal Forest Division.[19] This fact alone indicates the vast wealth of bamboo jungles which were more or less completely destroyed by the mid nineteen thirties under the so-called protection and scientific management of the Imperial Forest Department.

What had happened between 1868 and the turn of the century? The disappearance of this humble and oft ignored head of revenue from "minor forest produce", in fact, indicated the extinction of sales of all traditional sources of forest produce like skins, horns, produce of creepers, honey, lac, gum, tanning materials, fibers and cordage materials and a vast quanity of medicinal herbs. There were also other categories of extremely valuable forest produce which were not marketable and yet formed an integral part of the life of the people. Thus, in 1888, Hearle notes

> owing to the number of edible wild fruits, the forests are valuable to the starving people in times of famine and owing to the number of trees yielding fodder, give substantial support to cattle in times of drought.. un cultivated products are used as food chiefly by the inhabitants of the jungle tracts adjoining the Government forests and in times of scarcity by the people of the upper pattis. At all times, however, the young leaves of the nettle and several species of fern sorrel and the like are used as spinach by all classes.[20]

N.C. Hearle certainly stands head and shoulders above other British Forest Officers of the early period by virtue of his perceptive account of a large number of miscellaneous species other than sal and sein, as also by his acute and perceptive observations regarding the ecological relationships of various plant species. In many places he has noted indications of the death or extinction of particularly valuable species such as haldu (adina cordifolia). No tree was too humble to escape his notice. Detailed descriptions with age classes, habitat and reproduction trends were given for sal, sein, haldu, bakli, dhaura, gosam, shisham and khair, tendu, jamun, sandan. Kokat, which was the most numerous, contained all the other species amongst which the most valuable were bahera, semal, bel, padal and kaim. These were also mentioned by Dansey who dismissed others as valueless shrubs and inferior species. Not so Hearle who mentions ban, amaltas, saonjha, doldhak, dhak amongst the large trees and others such as aonla, pula, ber, riuni, chila and dudi which generally remained small trees. He further gives a three page list, which he does not claim to be exhaustive, of the most commonly found species, including shrubs and creepers.[21] Thus it is indeed interesting to see how British civil servants and forest officers progressed from the lively and extremely erudite accounts of miscellaneous forest species given by Atkinson and Hearle in the nineteenth century, to the utter indifference, if not ignorance, of Coombs who formulated the Working Plan of Lansdowne Forest Division in 1939-40 when minor forest produce brought in just Rs.270 per annum.

When the forest department started formulating working plans, these plans drew up a mandate to embark on a war against creepers and 'inferior species' which had still managed to survive the depredations of Company Raj! Thus the era of 'scientific management' saw a systematic eradication of creepers which were cut in thousands and thousands of acres in all the existing forest tracts then under the control of the department. Weeding out of 'inferior species' was also actively considered in the Garhwal forest division where it was tried out, initially in the North Patli Dun forests, and subsequently given up because it was not always possible to find purchasers for the timber (naturally, since the vast majority of species were valued not for timber but for renewable by-products). Further, so much debris and refuse could not be left lying around without fear of causing fire. The practice was, therefore, given up by Beadon Bryant in 1896 with the stipulation that these species should be sold off, even at reduced rates, where buyers could be found and when trees so marked were still left standing by the contractor, they were to be girdled and left to die a slow death. A regular war was waged against malhu (bauhinia vallhi) which used to serve multifarious needs of local people, its leaves in particular being used for manufacture of native leaf plates.[22] In all probability, this host of miscellaneous species provided leaf mould and compost that in turn built up a micro-environment suitable for nursing young sal. These multiple species must have helped to maintain an eco-system which provided the essential environment in which rich and healthy growth of the dominant species could be facilitated.

Such wanton destruction of all species other than the main commercial species of sal, sein etc. ultimately destroyed, not only the local economy, but also the eco-system which nurtured the sal forests themselves.

Though it is not a widely known fact, clear felling of forests for cultivation had been practiced even in pre-British days in Garhwal. In fact, Coombs, in the Working Plan of Lansdowne Division (1939-40) hints that over-felling of forests in the sal dominated tracts had preceded British rule and the native rulers should therefore share a portion of the blame for the state of the sal tract as it existed in the 1940's. Other writers have also tried to blame the Gurkha rule for destruction of the forests.[23] Certainly, it is an accepted fact that large portions of the northern slopes of the foothills forests of Garhwal had once been clear felled and brought under fairly extensive cultivation, some portions not less than twenty years before the British occupation and some dating back much earlier. E.P. Dansey in the 'Working Plan' for Patli Dun hill forests of 1896 mentions vast tracts which showed evidence of old cultivation of the Gurkha period (when many people fled from the hills) some dating to an even more ancient period. Udepur range, west of Kotri Dun was another example of ancient cultivation.[24]

How did this system of clear felling for cultivation (much maligned in recent times as 'jhum' cultivation) react on the eco-system? How was its impact different, if at all, from the system of clear felling followed by the British forest department?

The impact of ancient cultivation on the state of the forests may best be culled from the first Working Plan of Garhwal drawn up for the Patli Dun hill forests by E.P. Dansey. The latter notes that on all these tracts, some of which were felled and cultivated 60 years before,

> the areas have since been filled up with sal of the 3rd or 4th class only but instances are not uncommon in which sal of the 2nd or even 1st class form a thick and regular forest over the ancient sites.

Thus though areas in compartment No.1 of Division-III had been under cultivation at one time, it was described by Dansey as "altogether a magnificent bit of sal forest in superb condition". Again large portions appeared to have been under former cultivation in Compartment No.2 but on the southern aspects (of these portions)

> the forest has completely repossessed itself of sal... The northerly face however was found to have poor cover due to heavy damage of frost on seedlings... the cold was, in fact, not conducive to sal and in this compartment, along ravines, banj (Querius incana) was already making an appearance. Only some portions could not withstand the cold and these were marked by blanks where there had once been cultivation.

However, in general, the condition of the forests previously worked by the villagers was so good that Dansey could state quite categorically

> *While numerous tracts were denuded at a former epoch and placed under cultivation... all these areas for the most part filled up again with forests* and the only important evidences of working that we now meet with must be attributed to a more recent period when extraction by sal on the part of the lessees and later on by the Forest Department became recognised facts.[25]

Another testimony to the fact that ancient native cultivation did not destroy the eco-system, even though it was based on clear felling, is provided by Major Pearson while describing the Kumaon forests of the Kosillah river in 1868, where "exceedingly fine and regular young forests" were springing up as in Amtoonolh and Chukur Nagul... "almost as regular in appearance as in plantations" with trees upto 3 feet in girth and 50-60 feet in height. The Conservator marvelled at the fact that

> the whole of the plateau on which these forests stand bears the mark of ancient cultivation, the trees springing up on the edge of the old fields. This is said to date from the time of the Chund dynasty, about a century ago, *and one mature forest of trees has already been removed to make way for the forests which are now growing up.*[26]

To use a somewhat vulgar phrase, the proof of the pudding lies in the eating. Clear felling for cultivation in the indigenous system was yet followed by equally good regeneration whereas this was not so in the areas clear felled under British rule. Why was this so?

The reasons are not difficult to find. Clear felling under British rule had adverse environmental impact because it covered continuous vast tracts, sometimes in over a thousand hectares, whereas local people generally made small or compact and dispersed clearings. These continued to remain surrounded by forests and were fed by small streams with attendant beneficial effects on the soil derived from silt and compost. Secondly, even though clear felling for cultivation involved burning of refuse, this was a controlled and one - time affair which enriched the soil, rather than impoverishing it. Regular cultivation over a period of time also had soil binding effects and since the rotation of crops was based upon traditional knowledge of soil enrichening versus soil depleting crops, the overall effect of agriculture was not to render the soils impoverished (as happened under British rule due to repeated fires, destruction of soil cover and long exposure to the sun). The principles of forestry as evolved by the British in the nineteenth and twentieth centuries had, in fact, involved waging a regular war on the eco-system, and this happened not only in the Himalayan forests, but also in the great forest tracts of the Central Indian plateau in Madhya Pradesh.[27] These silvicultural practices were termed as scientific forestry management, perhaps because they largely suited the primary economic interests of the rulers.

Today the developed world has completely changed its stance. Having spearheaded ruthless commercial exploitation of forests for which they evolved so-called scientific principles, the former colonial powers are now anxious to stop commercial felling of tropical rain forests by Third World countries which have freed themselves of the colonial yoke. What was sauce for the goose is certainly not sauce for the gander.

The new slogan of the ex-colonial powers has now become conservation of biogenetic resources and eco-systems. In fact there have been distinct moves to have the forests of the developing countries declared as a "global resource", a gene bank which will then be regarded as a natural heritage area belonging to the whole world, instead of remaining within exclusive jurisdiction of nationalist, sovereign states.

While it would be uncharitable on the part of developing countries to see only forces of vested interests in this concern for conservation of genetic resources, it cannot escape attention that so-called scientific approaches to the environment on the part of western powers seem to be quite powerfully dictated by historical circumstances. Obviously they are influenced by dominant commercial interests of the particular period in which they are made.

NOTES

1. G.F. Pearson. "Sub-Himalayan forests of Kumaon and Garhwal" in *Selections from records of the Government of the North West Provinces*, 2nd series, Vol. 2, Allahabad 1869.
2. *Ibid.*
3. *Ibid.*
4. *Ibid.*
5. E.P. Dansey. *Working Plans for the Garhwal Forest division 1879-80.*
6. *Ibid.*
7. Coombs W. *Working Plan of Landsdowne Forest Division 1930* .
8. Osmaston A.E. *Working Plan of North Garhwal Forest Division, United Provinces, 1921-22 to 1930-31;* Government Press Allahabad; 1921, pp.110A, 112A.
9. Osmaston *op. cit.*, para 13; p.6.
10. A.N. Chaturvedi, 'Environmental Value of a forest in Almora' in *The Price of Forests* (ed) Anil Agarwal CSE 1990. Gives an excellent presentation of the economic value of broadleaf forests for maintaining and regulating water supply for human settlements, both rural and urban.
11. Richard St. Barbe Baker; *My Life My Trees.*
12. Gupta, P.N. 1979, *Afforestation, Integrated Watershed Management and Land use Development Project for the UP Himalayas and the Shiwaliks.*

13. Atkinson *Gazetteer* Vol. I; Part II; Economic Botany, pp. 672-844; Also see:
N.C. Hearle *Working Plan for the Ganges Working Cirlce, Ganges Forest Division, Central Circle, NWP and Oudh,* Allahabad; 1888.
Hearle also mentions *Economic Botany* as an interesting chapter of the *Gazetteer* which gives an account of different kinds of minor produce exported directly from the forests or brought through them usually via Kotdwar from the upper hills and states. To do justice to the vast array of species, would, according to Hearle, require a volume in itself. He gives a short list showing the edible fruits etc. and the principal products exported for use as medicine, dyes and tans, gums, resins and fibres. (pp. 39-41) He also draws attention to the large number of plants " "said to possess medicinal properties by the natives, *imaginary of course in some cases*".
14. Osmaston; *op. cit.* p. 27.
15. Atkinson *op. cit.*
16. N.C. Hearle *Working Plan for the Ganges Working Cirlce, Ganges Forest Division, Central Circle, NWP and Oudh,* Allahabad; 1888; Appendix B, List of minor produce and rates of royalty from 1885-86, pp.7a to 9a.
17. Walton, H.G., *British Garhwal: a Gazetteer;* Allahabad; 1910; Reprint Nataraj Publishers, Dehra Dun; 1989; p. 90.
18. Coombs *op. cit.*
19. Atkinson, *Gazetteer op. cit.*; pp. 851-862. Dietrich Brandis in 1880 recorded his opinion that in many places in Garhwal division the bamboos were being overcut. Also see Hearle *op. cit.* p.31; para 59. In para 64 the average revenue reciepts from bamboo in the combined Garhwal and Ganges division 1873-1883, are given as below :

Average of first five years	Rs. 57,000
Average of second five years	Rs. 48,000

The Inspector General of Forests concluded in 1883 that "the export of bamboos has fallen off during the last ten years and the revenue has been reduced accordingly."
20. Hearle *op. cit.*; para 77, p. 38.
21. Hearle *op. cit.*; pp. 39-41.
22. F.Beadon Bryant, *Working Plan of Garhwal Forest Division, Central Circle NWP and Oudh*; 1896; para 58, p.25; "Creeper Cuttings" states that in addition to the war waged against the destructive maljhan all creepers were being cut over such areas as are to be felled during the following years. A detailed table giving yearwise and compartmentwise lists indicates that a total area of 64,270 acres was cut over at an average cost of one anna one pie per acre.
23. Thomas Weber, *Hugging the Trees;* Penguin; 1989 Ch.1; p. 19.

24. Dansey, E.P. "*Working Plan of Garhwal, Patli Dun Hill Forests*"; *1896;* quoted in F. Beadon Bryant *op. cit.*; para 15, p. 7.
25. *Ibid.*
26. Pearson, G.F.; *op. cit.; Selections.*
27. Buch, M.N. *The Forest of Madhya Pradesh*; Madhya Pradesh Madhyam; Bhopal, 1991.

CHAPTER 4

Destruction of the Pastoral Economy Cattle vs Forests

Over a period of time, the revenue and forest departments of the Imperial Government appear to have struck a silent truce, by which the revenue department concentrated on realizing higher revenues for Government by way of expanding agricultural operations, while the forest department earned increasing revenues from clear felling of forests, from which men and cattle had to be excluded. Thus the village commons, already attenuated, came to cater to the expansionist needs of the revenue department, as also to the needs of cattle, which had to be increasingly confined to the areas outside the notified forests. It is not surprising, therefore, that cattle population became the first and major casualty of these policies. Let us, look at the process by which this came about in more detail, turning back to the formative years of the earliest working Plans of the Garhwal and Ganges Forest Divisions.

Complete exclusion of cattle was a tenet of forest management which had to follow, ipso-facto, upon the principle of clear felling, since the felled block had to be completely closed to human and cattle interference in order to allow the second class trees and the young seedlings to regenerate and grow into first class timber. Hence, it came about, that the hill people were systematically dispossessed of their cattle stations and, either completely driven out from some areas, or granted limited grazing rights (free of tax) in others which were generally the worst and most degraded blocks.

Earlier chapters have quoted at length from Major Pearson's historic report on the state of the sub-Himalayan tracts written in 1868 on the eve of their take over by the Imperial Forest Department. Some more extracts are given to stress the point that the forests were destroyed by the policies of the British rulers and exclusion of man and cattle followed as an inevitable consequence, making it impossible for the Garhwali to practice the system of rotational grazing and trans-humance on which success of the highland pastoral economy depended. Major Pearsons writes:

> In the Garhwal forests the Chandee forests may be said to have been completed worked out as their proximity to the Ganges has rendered them easily accessible to the contractors and agents of the Roorkee workshops and Public Works Department... tracts in the forests of Odeypore were considerably worked by Capt. Reid and now but little valuable timber remains in it. The forests in the valley of Ramganga, the South Patli Dun and Sonali may all be said to have been *felled even to desolation* by Capt. Reid as well as by Mr.Finn and the native contractors who went before him. Immediately above the cultivated lands of the Bhabur, west of Ramnagar the forests may be said to have been worked out and as they are required as grazing grounds for the cattle of the Bhabhar villages, nothing more can be done than to protect the Sal as a reserve tree for felling. In the Kotah Bhabur, some excellent Sal forests have been enough worked to require rest in the tract; there are several villages with cultivation near the hills *but their limits have been strictly defined and they are not allowed to keep any buffaloes.*

In 1858, the forests were taken over by Colonel Ramsay, Commissioner Kumaun, who strictly managed then for a decade. Felling of trees without permission was prohibited and Forest Officers were appointed.

> In 1861-62 cultivation in the Patli Dun was put a stop to by assigning land to the people below the hills in the Bhabur; next the herds of cattle stations were broken up and all herds of buffaloes removed from the forest.[1]

Hence by 1868, Major Pearson, Conservator of Forests, North West Province, could bear testimony to the satisfactory state of protection from *grazing and cattle menace,* congratulating Colonle Ramsay for the tact and efficiency with which all hereditary, time honored traditional cattle stations of local inhabitants had been displaced from their native habitat. It is clear from the above that, in the chronology of events, the forests on which the pastoral economy of the region depended were totally destroyed or 'worked out' (in forest department parlance) due to activities of the East India Company and its agents. *There was no cattle menance to the forests. The forest policies were a menace to the forests and sequentially to the cattle.*

While drawing up the Working Plan of North Patli Dun forests in 1881, E.P. Dansey mentions that the northern villages of the Mundhal valley grazed their extensive cattle herds in adjacent village forests, due to which their main occupation was animal husbandry rather than agriculture. Dansey suggests with true imperial sarcasm that agriculture is perhaps "too exacting an occupation for the indolent Garhwali who can keep large herds of cattle *costing him nothing to maintain*".[2] E.P. Dansey seems to have forgotten or been ignorant of the fact that animal husbandry has, traditionally, been the major occupation of highlanders all over the world, including his own country and Europe, simply because of the fact that where forest wealth is maintained for biomass needs of local inhabitants only (and not for commercial extraction,) maintenance of huge herds is possible and self-sustainable literally *at no cost* to the environment.

Fifteen years later, the position of both the forests and villagers of the Mundhal valley appeared to be less happy. F.Beadon Bryant, Deputy Conservator of Forests, drawing up the Garhwal Division Working Plan in 1896 mentions damage done by grazing in the immediate vicinity of the Mundhal valley in areas open to villagers' rights.[3]

What could have happened in the intervening period? The main occurrence which might have triggered off this deterioration was the Imperial notification of 1893, by which all lands not under cultivation were brought under Government control. The full implication of this notification for eco-system management in the Himalayas was far reaching in so far as it was based upon fallacious notions regarding non-profitability of fallow land, by declaring that all unclutivated lands vested in the state. This ultimately led to the extinction of the rich agricultural economy of Garhwal which was based on the traditional system of allowing lands to lie fallow so that they could be cultivated on the age old principle of rotation and rest.

Villagers who had continued to enjoy unlimited access to vast tracts of forest lands under traditional systems of management, were now driven out of the forests and forced to confine their cattle within specified boundaries where they were accorded limited 'rights' of grazing by the Forest Department. It is not surprising, therefore, that damage was being done by over grazing in these areas since, with one stroke of the pen, the British Government had ensured that incidence of grazing would have to be necessarily confined and intensified. The principle of trance humance of pasturage through a system of migration, rotation and rest, was completely abolished.

A study of the Bhabhur region can give an idea of the processes which led to destruction of the pastoral economy while simultaneously retarding regeneration of forests already depleted by the activities of the East India Company.

The Bhabhar was a strip of dense malarial forests which extended all along the foothills of Garhwal and Kumaun, in a 10-20 mile wide stretch that was specifically a "water-less tract". The region was called after bhabhar grass which was so extensively found there that there was enough, according to Atkinson, to keep the country provided for another century. The area abounded in all kinds of forest produce.[4] The 1910 Gazetteer of Garhwal clearly testifies that, with the exception of a few stray villages, the Bhabhar was never meant either for cultivation or for habitation, being in fact an extremely unhealthy tract of dense, damp, rich forests which yielded enormous quantity and variety of forest products and also provided vast pasturage for winter grazing of migratory cattle from Garhwal, Kumaun and adjoining plains areas. These cattle manured the soil, cropped rank grass and weeds close, thus permitting advance growth of young sal and other miscellaneous species which would otherwise have been choked up by uncontrolled growth of grass.

Since the Bhabhar was approachable only in winter, grazing of cattle and collection of forest products (subsequently termed 'minor' forest produce) used to be the major activity of local inhabitants during the annual winter migration.

Now another grievous blow was dealt to the pastoral economy by introduction of 'professional' graziers. In the indigenous system, the Bhabhar tracts had always belonged to hill people who, nevertheless, allowed graziers from adjoining plains of Rohilkhand to also graze their cattle there on payment of taxes known as Ghikar and Gaicharai. These taxes were fixed by the Raja of Garhwal and had been paid to the Garhwalis since many generations. In this system the village communities who owned various parts of the tract, apportioned and managed incidence of grazing amongst themselves, as also amongst the Rohilkhand tax paying graziers. Naturally the British had to destroy this system of traditional management of pastures. Though (as we will see during the course of this narrative) they could not ignore traditional rights altogether, in the Bhabhar tracts, they not only abolished the distinction between local inhabitants and outsiders, they went a step ahead by specifically discriminating against local people while giving preferential treatment to outsiders or professional graziers (as they were termed in the working plans) from whom the state directly realized taxes. That grazing taxes were quite considerable is borne out by the fact that in 1846-47 the Kumaun division obtained Rs.18,729 by way of forest dues, *half of which accrued from grazing taxes.*

Between 1865-1868 a total amount of Rs.28,649 was realized from grazing dues in the forests of Kumaun division. A detailed account of grazing fees being realized by the British Government in Garhwal can be found in the first working plan drawn up for Garhwal division in 1888 by N.Hearle, Assistant Conservator of Forests. These fees are mentioned at 8 annas per cow/bullock and 6 pies per sheep/goat.[5] In 1847 these dues had been 3 annas per buffalo and one anna per cow/bullock. The British Government had thus almost trebled its fees over a period of forty years. To the trusting reader, even the grazing fees of 1888 may not appear exorbitant, until the daily wage rate is highlighted. While Hearle makes no mention of daily wage rates, ten years later, Beadon Bryant, in the working plan for the adjacent Ganges division, mentions a daily wage rate of 2½ annas prevailing in Bhabhar. In his settlement report of 1896, Pauw has dwelt in detail upon inflationary trends in Garhwal.[6] However, to take a conservative estimate, the daily wage rate in 1888 may be assumed at one anna, though it could have been much less. This would mean that, for every cow grazed in the Bhabhar, the villager in 1888 was paying the equivalent of three day's daily wages. In case he owned twenty cows he would have to pay the equivalent of two months wage labour towards grazing fees. In the case of buffaloes the rate was even higher, being the equivalent of eight days' wages. In terms of prices prevailing in 1987, when this study was conducted, assuming an average daily wage rate of Rs.13 only, the tax works out to Rs.40 per head for cow/ bullocks and Rs.104 per head for buffaloes. It was to be expected that the British would preserve the best grasslands for this remunerative business which tended to get increasingly concentrated in the hands of big people or "professional exporters".

Hence it came about that the hill people who were the real owners of the land were systematically dispossessed and either driven out completely from certain areas or granted limited rights, free of grazing tax, in the worst and most degraded blocks. This can be proved by a detailed study of grazing rights and grazing areas allocated in the sub-Himalayan tracts of the Garhwal and Ganges Forest Divisions.

The Working Plan of Garhwal division (1896) divides the forests into three main classes.[7]

Forests of Class A were mainly forests covering an area of 40,000 acres in which "important quantities of sound Sal timber is still left". These forests were in prime condition and able to support grazing of cattle through sufficient biomass generation, but cattle and animal husbandry naturally did not figure anywhere in the list of priorities of the Forest Department. Hence it is not surprising to find the Working Plan confirms that "*more than half of this area was completely free of rights*" while the northern strip on the Mundhal side was open to a specified number of cattle of villagers only.

Forests of Class B formed an interesting category covering an area of over one lac acre and extending in an east-west strip that included huge open grasslands in South Patli Dun, in the vicinity of Ramnagar and Dhikala. The grassland in these blocks covered over 2000 acres, in a compact area extending along the left bank of the Ramganga river. These grasslands could have been given to the right holders or local inhabitants, most of them Kumaonis, who used to regularly bring their cattle down in large numbers to the bhabhar during the great writer migration to the foothills. However, this was not to be. The Kumaon cattle were confined to a narrow strip averaging one mile in width along the northern boundaries of the Kotah and Chilkia forest ranges. Maximum grazing facilities in excellent pasture range were, instead, given here to professional grazers on payment of dues which were not inconsiderable. At one point the working plan considered the possibility of attempting plantation in these open grasslands (which, incidentally were once thick sal jungle, now reduced to this state by the activities of Messers Reid and Finn) but the working plan concluded that grazing fees earned were also valuable, while the success of plantations would be doubtful.

Forests of Class C were the forests of the Malani and Laldhang ranges, along the southern most edge of the Garhwal forest division of the Bhabhar, covering an area of approximately 78000 acres, out of which an area of 19000 acres was open to rights for grazing by villagers. Since the Plan omits figures of the actual number of cattle allowed to graze in different areas, we do not have precise statistics to prove the assumption that these 19000 acres of Class C forests were opened to maximum number of cattle heads. However, this conclusion can be inferred from a number of other facts mentioned in the report. For instance, B. Bryant states that these forests *"are already over-burdened by rights"*, the incidence of grazing being apparently very high. Bhotia traders are mentioned as the main offenders. It is a well known fact that the Bhotias used to come down to the Bhabhar in winter with large herds of cattle along with various imported articles from Tibet. While the Working Plan is silent on this point, some salient figures could fortunately be culled from Pauw's invaluable Settlement Report of Garhwal which mentions the number of Bhotiya sheep and goats (which alone were taxable) as 9500 or 10000 in 1895-96 (just the year of the Working Plan). The Bhotias paid their dues of 6 pies per animal at Ramnagar (within Garhwal forest division) and in Kotdwar (within the adjacent Ganges forest division). Since Niti and Mana of Garhwal are relatively poor, while large and prosperous villages are found in Johar and Kali Byans valleys of Kumaon, one can tentatively suggest that Kumaon and Garhwal Bhotiyas together grazed not less than 15000 sheep and goats in the 19000 acres opened to them in the Garhwal forest division. It is not surprising that the forest officers complained most about this class who were, incidentally, people of considerable means and not to be easily bullied. While the incidence of grazing was, therefore, very high (perhaps an acre and a half per head of cattle), the condition of the forest itself was poor. It is a deplorable fact that such degraded forests were opened up to grazing and that too of a high incidence. Let the Working Plan speak for itself :

> These Class C forests (west of the Ramganga) are in a very unsatisfactory condition... they consist almost entirely of poor, open mixed forest where, as a rule very little good Sal is to be found, inadequately clothing the steep unstable hills of the outer Himalayas. Their protection is of the greatest importance in the interests of the cultivation lying below and outside them. The main object of management in these ranges was, therefore not so much the maintenance of the forest for commercial purposes but its protection in order to prevent landslips and disastrous floods and to ensure an increased and permanent water supply to the cultivators of the plains.[8]

How this objective could be achieved while leaving the area open to so many cattle is a question that begs an answer?

N.C. Hearle's Working Plan of the Ganges Forest Division 1888 (the forests of which were mentioned as being in a totally wretched condition), gives further details which reinforce the hypothesis that the British Government preserved its best grasslands for professional grazers, 'big people' who could afford to pay heavy grazing fees, while the ordinary villagers, whose native forests had been expropriated, were relegated (albeit free of grazing tax) to the worst and most confined areas.[9] The figures given in the Working Plan are as below :

Category of Cattle	*Nos.*	*Area grazed*	*Acres per head of cattle*
Right holders	13,288	18,106	1.36
On grazing fees	5,275	75,282	14.00

These figures speak for themselves and establish the fact of dispossession most clearly as far as Ganges Forest Division is concerned, though similar conclusions have been inferred from external sources in the preceding paragraphs for the Garhwal Forest Division, as well, covering roughly the same decade 1880-1890 or 1885-1896. In the Working Plan of the Ganges Forest Division, Hearle himself notes the disproportion in area availability to both categories of cattle but explains it away thus:

> The number of acres per head of cattle (for professional grazers) appear very high in some cases but it must be recollected that most of the ground is inaccessible and that most of the grazing is confined to the flat ground and to the more moderate slopes. The villagers have forests and wastelands of their own besides the Government forests which are in consequence not grazed over by village cattle to the full extent permissible.

Like other British Officers, Hearle also found it convenient to forget that the rich pastoral economy of Garhwal had been based on a careful system of eco-management through trans humance where village grounds were meant to be rested by rotation. Only five blocks had been opened to the right holders as against ten blocks for the professional grazers. Not only did the villagers cattle have to make do per head with 1.46 acres of land which included slopes, ravines etc. (the net available area being much less), but this land was itself also in an extremely degraded condition as is borne out by the blockwise descriptions of the forest and the stipulation that the trees here may be sold "even at reduced rates".[10]

The grazing situation as described above was certainly a complicated one. Despite the high incidence of grazing in the four blocks opened to villagers with recorded rights, Hearle could still state categorically that the division may be considered to be overall free from undue harm from grazing. Despite all these assertions in three working plans of the preceding decade, Walton, in the District Gazetteer of 1910 states peevishly that

> grazing dues are small while parts of the forest in Garhwal forest division are heavily grazed and in such parts the crop is very open and regeneration almost nil.[11]

An analytical study of the divisional working plans has shown, however, that in most cases, areas which were already desolate and degraded were handed over for grazing to a large number of right holders cattle. The latter could not be expected to get much sustenance out of these and the already degraded forests could hardly rejuvenate. Right from Walton down to the present day, however, policy makers and administrators have widely believed that cattle population is the cause of degradation of forests.

Were the Himalayan forests really destroyed by cattle? Or were the forests destroyed in the first instance by reckless commercial exploitation, the residue being left to hapless animals who had to make do somehow and also bear the curses of subsequent generations? Curses fortunately go unnoticed in their case, since with cattle, unlike with man, you cannot add insult to injury. Cattle being dumb can only give up the struggle in silence, reducing in health, in milk yields and finally in numbers.

In a perceptive paper on the "Role of Leaf Fodder in the United Provinces" M.D. Chaturvedi, B.Sc.(Oxon) IFS, Chief Conservator of Forests, Uttar Pradesh, writing in 1948, examines "the truth of the oft repeated assertion that the bovine (cows and buffaloes) population is far in excess". He correlates the cattle stock with the population and area under cultivation in different natural regions in the United Provinces and concludes as follows:

> From these figures emerges the startling fact that far from being in excess, the bovine stock is hardly sufficient to cope with the agricultural requirements of these provinces. Thus there is one work animal for every 6.2 acres of cultivation in the Himalayan tract. 3.25 acres in the Gangetic basin and 4.8 acres in the Central

> Indian Plateau. This is far from being sufficient, more particularly if the inefficiency of the ill-breed cattle of this region is remembered. The position of milch and breeding animals is even worse. There is only one cow or buffalo for nine persons in the Himalayan tract, six persons in the Gangetic basin and for three in the Central India Plateau.[12]

The overall cattle population of the United Provinces had gone down from 22,296 in 1920 to 21,742 in 1945 but the decline in the Himalayan tract would have been greater.

In 1885, Hunter could report in the Imperial Gazetteer that

> people have grown rich in later years.. they keep more cattle and get more manure.. rice and mandua are in surplus.

In 1869 when famine conditions prevailed in the adjacent Bijnore district, Garhwali peasants earned handsome profit by way of vast export of grain to the scarcity hit areas. The surplus was documented till 1881 and was being exported both to Bijnore and Tibet. Peasants paid their revenue in cash and indeed it was one of the few districts where revenue could be collected with such ease. Even in 1910 the gazeteer of Garhwal noted that "miscellanous earnings of hillman were high" through sale of ghee, manufacture of woollen goods and load carriage by ponies and mountain goats.

> The people on the whole feed better. One rarely, if ever, sees a beggar. The houses are large, solid and comfortable, and the great majority of people are well dressed.

History has shown that results of ecological degradation tend to be adversely experienced only after a space of half a century or so. In the case of Garhwal the tremendous bounty of nature somehow continued to sustain the people over the first 60-70 years of British rule. Even after fifty years of depredation caused by the Company Raj the villagers continued to prosper on the residual bounty. But by the turn of the twentieth century the wheel had started to come full circle. By then there was no cream left in the milk, the platter was already expropriated and only some dregs in the form of rights and concessions were left to the people. From this period starts the occurence of successive famines and the slow steady decline of both agriculture and the pastoral economy, leading to the ultimate slow death of the U.P. Himalayas.[13]

NOTES

1. G.F. Pearson. *"Sub-Himalayan forests of Kumaon and Garhwal"* in Selections from records of the Government of the North West Provinces, 2nd series, Vol. 2, Allahabad 1869.
2. E.P. Dansey. *Working Plan for the Patli Dun Hill Forests,* 1881.
3. F. Beadon Bryants. *Working Plan of Garhwal Forest Division, Central Cirlce;* NW Provinces Oudh, 1896; p.18, para 45.
4. E.T. Atkinson. *The Himalayan Gazetteer;* Vol. III, Part I; Bhabhar; pp. 50-74 gives a detailed account of the region and the process of how it was colonised and cultivation started.
5. N.C. Hearle. *Working Plan for the Ganges Forest Division,* 1896 pp. 43-44.
6. F. Beadon Bryant. *op. cit.*; p.31.
   Pauw, E.K. *Report on the Tenth Settlement of the Garhwal District;* Allahabad Govt. Press; 1896. Chapter IV, para 70 gives details of inflationary trends. Paras 122-123 can also be consulted for the early history of cultivation and assessment in the Bhabhar.
7. F. Beadon Bryant. *op. cit.*; pp. 32-33.
8. *Ibid.* para 74, pp.33, 53-55; para 130, p.55 on "Grazing and other Rights" is of particular interest.
9. N.C. Hearle. *op. cit.;* pp.43-44; The figures quoted in the text are from the table in para 84 p.44.
10. *Ibid.*
11. H.G. Walton, *British Garhwal: a Gazetteer;* Allahabad; 1910; Reprint Nataraj Publishers, Dehra Dun; 1989; p.15.
12. Chaturvedi, M.D.; Chief Conservator of Forests, Uttar Pradesh; *Role of Leaf Fodder in the United Provinces;* Govt. Press, Allahabad; 1948.
13. E.K. Pauw. *op. cit.:* See Chapter II para 71 for scarcity conditions in 1877-78, 1890, 1892, and in the northern parganas 1893 and 1894. These were unknown under British Rule before the last settlement, as per Pauw's own statement. The succeeding paras of this chapter describe the adverse effect on public health, the increase in death rate and decline in the birth rate. He clearly mentions that the change from plenty to famine had started along with the beginning of a pauper class which had been non-existent before the advent of British rule.

CHAPTER 5

Destruction Decade in U.P. Himalayas, 1970-80

From the turn of the twentieth century we now take a quantum jump into free India in the decade of the seventies. The Indian government continued to manage the Himalayan forests through the Forest Department, manned by Indian officers, who functioned largely according to prescriptions and policies of forest management laid down by the British.

Why the decade of the seventies, one may ask? A practical reason is that it was the decade immediately preceding the period of the writer's study and statistics were also readily available in the latest publication of the Forest Department of U.P. *viz.* Forest Statistics of U.P., 1983. But the decade of the seventies was also important for other reasons. This period saw a sudden upsurge of consumerism, a spurt in easy money which led to increasing pull of urban markets and strengthening of the forces which go to support and promote reckless commercial exploitation of timber wealth. In the great forests of Central India, particularly in Madhya Pradesh, it was the years immediately following independence which witnessed wholesale decimation of forests, for after the ill-conceived policy of taking over Malguzars forests and other private forests by the Forest Department there was no incentive left to the people and the Malguzar to preserve the forest wealth which was no longer to be considered as their own. Thus large tracts of forests in Cental India were denuded and also lacs of hectares encroached upon by land hungry poor peasants.[1]

The British had encouraged the Garhwali peasant to extend cultivation at the expense of forests for the sake of earning extra revenue. In the forests of Central India the politicians of free India similarly encouraged the peasants for the sake of earning extra votes. In the U.P. Himalayas, however, there was little scope left for politicians to earn extra votes in this manner. No doubt, there was some reckless decimation of oak belts to make way for apple orchards and fruit belts after independence, but the major thrust of deforestation was occasioned by the greedy demands and the great pull of large and flourishing urban markets. Forest revenues were an attraction to governments perennially short of cash. The spirit of prosperity in other parts of India increased urban demand and accelerated the pace of commercial exploitation. The Chinese invasion of 1962 led to expansion and improvement of roads and this also provided a boost to the transport of produce for urban markets.

TABLE 1

Forest Area in Plains and Hills of U.P. (1981-82)

(Area in Square Kilometres)

Region	*Geog. Area*	*Population*	*Forest Area*	*Per Capita Forest Area*	*%Forest Area to Geog. Area*	*Area under Forest Deptt.*	*%age of Col. 6 to Col. 3*
	1	*2*	*3*	*4*	*5*	*6*	*7*
Hills	51125	4836000	34399	0.711	67.3	23819	69.25
Plains	243288	106026000	16867	0.016	6.9	16867	100.00
Total	294413	110862000	51266	0.046	17.4	40687	79.36

It is perhaps no coincidence that the decade of the seventies saw the culmination of the Chipko movement which was later to gain world wide fame. The rapid pace of commercial exploitation, coming as it did after almost 150 years of continuous decimation of forests conducted under the banner of scientific 'management', had to trigger off an equally strong reaction. Freedom and self-governance did finally yield their fruits for the people of Garhwal. The Chipko movement, no doubt, achieved some success. A ten year moratorium on cutting of green trees was imposed by Prime Minister Indira Gandhi in the early eighties.[2] Officially, atleast, commercial extraction ceased and the pendulum swung the other way in a desperate rush towards investments and increased expenditure for greening the Himalayas in the next decade. But before turning to that period, the records of the forest department of U.P. pertaining to the decade of the seventies merit indepth analysis.

There is a rough parity between the forest department area in the hills and plains districts of Uttar Pradesh, in the sense that 58 per cent of both the department area, as well as the exploitable forest area, fall in the hill districts. Keeping in view the National Forest Policy injunction to maintain 60 per cent forest cover in the hills, and about half that in the plains, one would expect the burden of timber and fuel production to be equally shared between the two; infact the plains districts could have borne a higher proportion of the burden. The figures, however, show a reverse trend. Between 1970 and 1981 the hills of Uttarakhand yielded about 63 per cent of the sawn timber of Uttar Pradesh as can be seen below:

TABLE 2

Outturn of Major Forest Produce in U.P. and Hills

Year	*Timber (Sawn) Cum*		*Fuel Cum*	
	U.P.	*Hills*	*U.P.*	*Hills*
1970-71	991207	627745 (63.3)	2478045	1857365 (74.9)
1971-72	1042651	595115 (57.07)	2315275	1548230 (66.8)
1972-73	956145	532245 (60.89)	2250563	1560479 (69.33)
1973-74	823017	475933 (57.82)	2583877	1854292 (73.32)
1974-75	68365	6412382 (60.32)	2126653	1315228 (61.84)
1975-76	874147	583463 (66.74)	2278427	108839 (48.66)
1976-77	864549	538386 (62.27)	2091742	1398483 66.85)
1977-78	858745	530557 (61.78)	2456553	1831745 (74.56)

1978-79	832926	565557 (67.9)	2355274	1733427 (73.61)
1979-80	682835	402866 (58.99)	2359103	1672236 (70.88)
1980-81	572427	352250 (61.50)	1861767	1353727 (72.70)
1981-82	655365	400762 (61.15)	1749806	1368436 (78.20)

(Figures in parentheses denote percentage values).

This contribution went upto about 67 per cent between 1975 and 1976, and 68 per cent between 1978 and 1979. The average annual production during the decade was about 5 lakh cubic metres of sawn timber and 15.5 lakh cubic metres of fuel. The actual output of timber in the round would be much higher, since wastage in the hills is considerable (about 30 per cent). In lay terms, we may safely state, that out of every ten trees felled for timber during this decade by the U.P. Forest Department, at least seven were cut from Uttarakhand.

TABLE 3

Regionwise composition of Forest Area under the Control of Forest Department

(Area in Hectares)

Species	*Forest Area in U.P.*	*% Forest Area by Species*	*Forest area in U.P.Hills*	*%Forest Area by Species in Hill Region*
Chir	411782	10.12	411782	17.29
Deodar	18411	0.45	18411	0.77
Firs Spruce	89405	2.20	89405	3.75
Kail	18417	0.45	18417	0.77
Cypress	3295	0.08	3205	0.14
Sal	535287	13.16	210534	8.84
Teak	51792	1.27	3205	0.13
Khair & Sissoo	74429	1.83	25091	1.05
Blanks	958120	23.55	602422	25.29
Miscellaneous	1907780	46.89	999403	41.96
Total	4068718	100.00	2381965	100.00

The main output of timber is limited in the hills to chir pine, kail (blue pine), fir, spruce, cypress and deodar. These species together occupy only 23 per cent (5,413 sq km) of the total forest department area of 23,819 sq km. Actually this area carries the main human settlements lying between an altitude of 700m and 2,500m and is under heavy pressure for meeting biomass needs and the demands of pilgrim and tourist traffic. Unfortunately, these coniferous forest belts also include significant tracts of oak which are the sole remaining guardians of the of the lower and upper Ganga catchments and are under acute stress, since broad leaved forests have been systematicllay destroyed by the forest policies laid down by the British and never reversed by independent India.

Further cause for concern arises when the extraction figures for the different species are studied. While sal (shorea robusta) outputs had peaked by 1970-71 and declined thereafter (no doubt due to over exploitation), the output of chir and other coniferous species went up by leaps and bounds. Thus, from an output of 2 lakh cubic metres in 1970-71, chir extraction went up to 3.19 lakh cubic metres in 1978-79 (sawn wood). The output for fir and spruce went up in the same period from 40,000 to 85,000 cubic metres. Deodar, the most noble tree of the Himalayas, was also cut heavily during this decade. Within a single year, (from 1971 to 1972) its extraction shot up by 480 per cent and in 1981-82 it was only marginally lower than 1970-71 outputs.

As the forest wealth of the foothills region of terai became depleted by 1971, the march of commercial exploitation reached higher and higher into the hills. It is not surprising that a study of satellite imagery of landsatt I (1972) made by the U.P. Land Survey Directorate showed the severest erosion intensity in the sal dominated Siwalik region.[3]

A similar process is very likely to have taken place in the Bhagirathi and Alaknanda catchments by 1981 but detailed studies in this direction have yet to be undertaken. Meanwhile the devastating and unprecedented floods of 1977 and 1978 are a grim and terrible reminder of the old legend—when the waters crashed upon the earth without the protective cover of Shiva's locks, they literally carried away the earth with them. Thus, while the floods in the country between 1960 and 1979 peaked in 1970, during the next decade these figures more than doubled, with the 1978 floods affecting 190 lac hectares of area and 700 lakh people (70 per cent of the flood-prone area of India falls within the Indo-Gangetic plain).

And all this to what end? To gain a net surplus revenue from the hills at an average of Rs. 8.3 crore a year though, in absolute terms, the surplus rose from 8.3 crore in 1970-71 to Rs. 16.34 crore in 1981-82 (against a gross revenue from the hills of Rs. 12.21 crore and Rs. 32.81 crore respectively). This is evident from figures given below:

TABLE 4

Expenditure/ Revenue Trends

Year	*Gross Total Revenue*		*Expenditure*	
	U.P	*Hills*	*U.P.*	*Hills*
1970-71	2148	1221(56.94)	757	389(51.4)
1971-72	2182	1232(56.45)	830	388(46.7)
1972-73	2466	1312(53.2)	952	454(47.7)
1973-74	2737	1501(54.87)	1109	513(46.2)
1974-75	3795	1945(51.2)	1360	767(56.4)
1975-76	3685	1785(48.4)	1526	854(56)
1976-77	3904	1802(46.1)	1642	947(57.7)
1977-78	3917	1925(49)	1803	920(51)
1978-79	4226	2020(47.8)	2050	1081(52.7)
1979-80	4518	2337(51.7)	2413	1223(50.6)
1980-81	4913	2614(51.7)	2413	1223(50.6)
1981-82	5748	3281(57)	3634	1647(45.3)

TABLE 4A

Net Surplus

Year	*U.P*	*Hills*	*Year*	*U.P.*	*Hills*
1970-71	1391	832(60.0)	76-77	2262	858(37.8)
1971-72	1353	844(62.0)	77-78	2144	1005(47.5)
1972-73	1514	858(56.6)	78-79	2176	939(43.1)
1973-74	1628	988(60.7)	79-80	2105	1114(52.9)
1974-75	2436	1178(48.3)	80-81	1748	1191(68.0)
1975-76	2160	931(41.0)	81-82	2113	1634(77.0)

It is axiomatic that the revenue forests or village commons have now been completely depleted with the excessive burden they have had to bear over the last eight decades, but not *only are the local people not being given enhanced rights in the remaining reserve forest area, from the figures it appears that these rights are being further curtailed.* Thus in 1977-78 the right holders were issued 42,000 cubic metres of timber as against 33,000 cubic metres issued to them three years later—a decline of about 8 per cent. Their share in the overall timber output also further declined from 6.4 per cent to 5.6 per cent while the purchaser's portion, the lion's share, went up from 88 per cent to 93 per cent. The next table bears this out.

TABLE 5

Outturn of Timber in the Round Agencywise Consumption/ Purchase

(cu.m.)

Agency	*1981-82*		*1980-81*		*1979-80*	
	U.P.	*Hills*	*U.P.*	*Hills*	*U.P.*	*Hills*
Deptt.	13867	6683	8987	5856	19519	6683
	(1.46)	(1.14)	(1.13)	(1.01)	(2.01)	(1.02)
Purch-	878750	527690	737998	521304	881445	579880
asers	(92.78)	(89.75)	(92.69)	(90.55)	(90.65)	(88.47)
Free	21413	20486	32632	15941	28982	26901
grantee	(2.27)	(3.48)	(4.10)	(2.77)	(2.98)	(4.10)
Right	33082	33082	16598	32609	42417	42018
holders	(3.49)	(5.63)	(2.08)	(5.67)	(4.36)	(6.41)
Total	947112	587941	796215	575710	972363	655482
	(100.0)	(100.0)	(100.0)	(100.0)	(100.0)	(100.0)

Since policy planners tend to aim all their prescriptions and remedies at the inhabitants of this region, they must consider them responsible for a destruction process which the forest statistics of Uttar Pradesh, however, show to be primarily in the hands of commercial and industrial interests. True, there has been a ban on felling of green trees since S.L.Bahuguna's historic fast of 1981. But the figures pertaining to the 1981-86 period which are analysed in the next chapter do not give much cause for optimism. Resin tapping continues unabated and felling of trees by storms has, infact, even exceeded the felling as prescribed in the working plans. Meanwhile the last decade may, in the light of the figures given above, be easily termed 'Devastation Decade' in the U.P. Himalayas.

TABLE 6

Fuel Outturn Given to Right Holders and Others (commercial purchase etc.) in U.P. Hills

(Outturn in cu. m.)

Year	*Total Outturn*	*Freegrants and right holders etc.*	*Purchases Dept. etc.*	*% of Col.3 to Col.2*
1974-75	1315228	655797	659431	49.86
1975-76	1108839	538141	5706198	48.53
1976-77	1139483	726071	413412	51.91
1977-78	1831745	717750	1113995	39.18
1978-79	1733427	756782	976645	43.65
1979-80	1672236	745416	926820	44.57
1980-81	1353727	753660	600067	55.67
1981-82	1368436	668715	699721	48.87
% variation during above period	4.05	1.97	6.12	

S.P. Mukherji, as Chairman of the Working Group on Integrated Action Plan for flood control in the Indo-Gangetic plains, December 1978, had recommended as follows :

> The Himalayas play a decisive role in the river regime in the Gangetic plain. I would have liked to recommend a complete moratorium on the felling of trees in the entire Himalayan catchment for the next 5 to 10 years and restoring the step hillsides to where they belong nature.... Atleast the Himalayan catchment in Himachal and Uttar Pradesh should be brought immediately under protection forestry, subject to the minimum needs of the local population being met.

Such recommendations are rare, though at one point dedicated officers in the Environment Department of the Government of India had lobbied in vain for such a ban. How is it that a democratic country not only curtails the fundamental rights of a large chunk of its citizens in order to cater to the needs of a burgeoning urban and industrial estate, but then also holds these very citizens responsible for the subsequent destruction? Why is insult added to injury? Why is the thrust of our research and policy prescriptions not aimed against these purchasers and consumers of 90-95 per cent of the timber and over 50 per cent of the fuel wealth extracted from the Uttar Pradesh Himalayas?

NOTES

1. M.N. Buch. *The Forests of Madhya Pradesh;* Madhya Pradesh Madhyam, Bhopal; P.O. Box No. 325; 1991; Chap. IV, pp. 41, 42, 46, 47, 61.
2. Anupa Mishra, Satyendra Tripathi. Chipko Movement; *People's Actions for Development and Justice,* December 1978. This monograph gives a good account of the involvement of Planning Commission and State Government Officials with Chandi Prasad Bhatt, and of the administrative and political moves which finally led to the imposition of the ten year ban on green felling in the U.P. Himalayas.
3. P.N. Gupta; 1979. *Afforestation, Integrated Watershed Management and Land Use Development Project for the U.P. Himalayas and the Siwaliks.* The tabular data also figures in:
   Neeru Nanda, "Who is destroying the Himalayan forests?" in *Himalayan Environment and Culture* (ed) N. K. Rustomji and Charles Ramble Indian Institute of Advanced Studies, Shimla, 1990.
   All the statistics quoted in this chapter are drawn from *The Forest Statistics of U.P. 1981-82,* Forest Department of Uttar Pradesh 1983.
   For an impressive and exhaustive analysis of state subsidy of forest produce supplied to industry, see the following articles by Madhava Gadgil, Centre for Ecological Studies Indian Institute of Science, Bangalore.
   Madhava Gadgil. "Deforestation : Problems and Prospects" in *History of Forestry in India,* (ed.) Ajay Singh Rawat, Indus Publishing Company, New Delhi, 1991.
   Madhava Gadgil. "Perverse Incentives for Industrialisation and its Impact on Forest Resources"; in *The Price of Forests* (ed.) Anil Agarwal; New Delhi Centre for Science and Environment; 1992.

CHAPTER 6

The Decline of the Reserves A Profile of Tehri Reserve Forests 1940-1982

Year after year the floods in the North of India seems to be getting worse, and this year they have been absolutely devastating. This means that there is something radically wrong in the Himalaya, and that "something" is, without doubt, connected with the forests. It is not, I believe, just a matter of deforestation as some people think, but largely a matter of change of species.

Living in Himalayas as I have been, continuosly, now for several years, I have become painfully aware of a vital change in species of trees which is creeping up and up the southern slopes, those very slopes which let down the flood waters on the plains below. This deadly changeover is from Banj (Himalaya Oak) to Chil (sic) pine. It is going on at an alarming speed and because it is not a matter of deforestation, but of change from one kind of forest to another, it is not being taken sufficiently seriously. In fact the quasi-commercial forest department is inclined to shut its eyes to the phenomenon, because the Banj brings them in no cash for the coffers, whereas the Chil pine is very profitable, yielding as it does both timber and resin.

But what are these crores of revenue worth if all, and more than all of that amount of money, has to be spent each year on trying to repair the damage caused by floods? I say "trying to repair", because year by year destruction gets worse, and takes many forms which are incalculable in terms of money.

Let me now explain in more detail and with the help of some illustrations, why the Banj is so much better than the Chil pine for holding back the waters of the monsoon rains.

The Banj leaves, falling as they do, year by year, create a rich black mould in which develops a thick tangled mass of undergrowth (bushes, creepers, and grasses) which in their turn add to the leaf mould deposit and the final result is a forest in which almost all the rain water becomes absorbed. Some of it evaporates back into the air and the rest percolates slowly down to the lower altitudes,

> giving out here and there beautiful sweet and cool springs. It would be difficult to imagine a more ideal shock-absorber for the monsoon rains than a Banj forest.
>
> The Chil pine produces just the opposite effect. It creates with its pine needless a smooth, dry carpet, which absorbs nothing, and which at the same time prevents the development of any undergrowth worth the name. In fact, often the ground in Chil pine forest is as bare as a desert. When the torrential rains of the monsoon beat down on these southern slopes of the Himalaya, much of the pine-needle carpet gets washed away with the water and erosion invariably takes place, because these needles, being non-absorbent, create no leaf-mould, but only a little very inferior soil, which is easily washed out from the rocks and stones.
>
> One very important point must be remembered in this respect, and that is when monsoon rain waters rush down and create floods, they run to waste and are lost to the cultivator. Whereas when they pass down slowly deep into the soil, they replenish the springs and river sources which are of such vital importance to agriculture in the dry season.
>
> But why are the Banj forests disappearing so fast ? It is not merely that the forest department spreads the Chil pine, but largely because the department does not seriously organise and control the lopping of the Banj trees by the villagers for cattle fodder, and as I have mentioned, is glad enough from the financial point of view to see the Banj dying out and the Chil pine taking its place. When the Banj trees get weak and scraggy, from overlopping, the Chil pine gets a footing in the forest, and once it grows up and starts casting its pine needless on the ground, all other trees die out.[1]

The above prophetic words of Meera Behn, written in 1950, state in a nutshell the problem facing the reserve forests of the U.P Himalayas with special reference to the district of Tehri Garhwal where she resided for more than a decade and a half. The diagnosis made by her is amply borne out by the data presented in this chapter which is based upon a study undertaken by the writer in 1987 in which a detailed comparison was made of the working plan of Tehri forest division drawn up in 1982-83 by A.M. Rai with the working plan of Tehri Forest Division drawn up in 1939-40 by P.N. Khosla. As a result of this exercise reliable statistics are now available to show the extent of degradation and deterioration which has occurred with reference to almost the entire forest area.

TABLE 1

Tehri Forest Divisional Area Statement by Main Forest Types 1939-40

Forest Type	*Area*		*Variation*	
	1939-40	*1982-83*	*Area*	*As %age of Col.2*
Chir Sub-Tropical Zone	52,000	57,000	5000(+)	10(+)
Himalayan Moist Temp. forests of Oak, Banj (Q Incana) & Morhu (Q Dilatata)	62,900	50,300	12600(-)	20(-)
Kharsu Oak (Q Semicarpifolia)	37,600	26,500	11100(-)	30(+)
All Oak Forests	1,00,500	76,800	23700(-)	24(-)
Oak Scrub	Nil	14,500	14500(+)	NA
Rocky, blank & snow	36,000	39,300	3300(+)	9(-)
Total Forest Area	2,46,000	2,34,000	12000	5(-)

Source : *Working Plan, 1939-40 P.N. Khosla and Working Plan, 1982-83 A.M. Rai.*

The Decline of the Oak Forests

In about 40 years the chir areas has increased from 52,000 hectares to 57,000 hectares *i.e.,* by 9.6 per cent. The oak and other deciduous forests however, have reduced from 1,00,500 hectares to 76,800 hectares *i.e.,* by 23.5 per cent. Even this does not give the full picture because, whereas no oak scrub existed in 1939-40, in 1982-83 14,500 hectares of oak had been reduced to scrub. As is well known, oak is not an invasive but a climax species and, therefore, the presence of oak scrub indicates that even in areas where oak still existed in 1982, excessive lopping had reduced the trees to mere poles or stunted bushes. If oak scrub area is added to the area where oak has totally disappeared, it is seen that 38 per cent of the oak area which existed in 1939 had been totally degraded by 1982.

It is a real cause for alarm that Tehri is losing about 600 hectares of oak forest every year in a process which has been going on for the last 40 years. The ban on cutting of green trees, imposed in 1981, had not stopped the felling of oak in U.P. hills, because in some parts of Garhwal oak forests were being worked for charcoal supply. Fortunately the ban operated in Tehri district and with charcoal manufacture having ceased, to that extent the oak forest recieved a slight reprieve.

The cause for concern is not confined to mere reduction of area under oak and increase in the area of chir. The density of both chir and oak has also substantially reduced, as can be seen even from an eye survey. To take the example of oak alone, a comparison of the oak working circle in 1939-40 and 1982-83 makes sad reading. In 1939-40 the profile of the oak working circle was as below:

TABLE 2

Oak Working Circle 1939-40

Area in Ha.	*Banj Oak (Q.Incana)*	*Chir Species*	*Other*	*Unworkable*
44,311	41,275 (93)	896 (2)	568 (1.3)	1492 (3.4)

TABLE 3

Increase of Unworkable Area under Oak Lopping Circle 1939-40 to 1982-83

Range	*Area of Oak W.C. 1939-40*	*Area under Oak 1939-40*	*Unworkable Area*		*%age increase of Unworkable Area*
			39-40	*82-83*	
1	2	3	4	5	6
Tehri	4,466 (10)	3,781 (85)	685 (14.5)	2,079 (55)	378
Bhilangana	16,085 (36)	15,535 (97)	168 (1)	3,103 (20)	1,747
Alaknanda	11,736 (26.5)	11,417 (97)	146 (1)	2,291 (20)	1,469
Maniknath	7,543 (17)	7.010 (93)	150 (2)	2,366 (34)	1,577

Note : Figures in brackets Col. 2 denote % of total area of Oak W.C. In other columns they denote % of area of range given in Col. 2.

In 1982-83 the Working Plan Officer made an enumeration of the oak areas which were by then unworkable because of heavy lopping, biotic pressure etc. The table gives a comparative statement of the increase in unworkable areas between 1939-40 and 1982-83 in forest ranges of Tehri.

In Tehri range the unworkable area has increased from 14.5 per cent of the total oak area of the working circle to 55 per cent, in Bhilangana from 1 per cent to 20 per cent, in Alaknanda from 1 per cent to 20 per cent and in Maniknath from 2 per cent to 34 per cent. The deciduous temperature zone, broad leafed oak forests represent the highest succession of vegetation in the Himalayan region. The above figures indicate the rapid rate of destruction of these forests which can lead to major environmental and ecological damage to the water management system, not only of the Himalayas but also of the Gangetic plain.

Decline of the Chir Forests

The position of chir is hardly any better. The chir forests of Tehri had always been one of the most striking features of the landscape. In 1868 Major Pearson, then Conservator of Forests of the North Western Province of Oudh, had described the route from Seansu in Tehri to Bhatwari in Uttarkashi as extending through a veritable sea of chir-pine forests. The quality of the chir forests can be gauged by the fact that, even in 1939-40, the unworkable area of the regular chir working circle was only 2 per cent of the total, as is evident from the table below :

TABLE 4

Chir Working Circle 1939-40

Total Area (in Acres)	*Unworkable (Unproductive)*	*Banj Oak*	*Chir Workables*
1,30,637	2,014	10,427	1,10,299

The major destruction of the chir forests would have started from 1951 onwards after accelerated pace of extraction described at length in the preceeding chapter. The results can be seen in the figures provided by the working plan of 1982-83, when almost one-third of the total chir area had to be consigned to the protection working circle because it had been depleted almost to the point of extinction. Thus, even though the area under Chir pine had increased in absolute terms, qualitatively there was a sharp decline in the crown density of chir forests.

Deforestation and Its Effect on the Climate

The table below gives the temperature charts in high summer at Tehri :

TABLE 5

Record of Maximum Temperatures in Tehri for the Month of May, June, July 1962 - 1980

Year	*May*	*June*	*July*
1962	36.7	37.6	34.5
1969	34.5	38.2	34.5
1975	41.3	42.2	NA
1979	39.9	41.9	39.5
1980	41.3	NA	35.9

(In Celcius Degrees)

TABLE 6

Average of 5 Years Maximum Temperature in Tehri for the Month of May and June

Year	*Average Recorded*	
	May	*June*
1962-69	36.00	37.56
1975-79	40.26	41.52

(**Source** : A.M. Rai's Working Plan of Tehri 1982-83)

The decade of the seventies was the period of maximum exploitation of the forests and there is clear evidence of the rise in summer temperature following over exploitation in Tehri. It is commonly observed that the hill areas are now much warmer than before, with people in Mussorie having to resort to the use of electric fans in 1987. The above tables almost conclusively prove that the higher summer temperatures are a result of massive destruction of vegetation cover in Tehri. Regardless of any global green house effect, deforestation seems to have created such conditions in the hills of western Uttar Pradesh.

Factors and Causes of Degradation

There are many factors leading to depletion of forest wealth and degradation of forest areas and unfortunately all of them seem to have been working simultaneously in Tehri.

Population increase cannot be held responsible. The decadal increase in population was minimal, being well below the national average. It increased only in 1971-81 and that too, in all likelihood, because of increased activity due to the Tehri dam.

TABLE 7

Population Growth of Tehri Forest Division 1939-40 to 1982-83

Year	*Households*	*Population Decadal*	*Increase %age*
1939-40	34,084	1,70,420	—
1951	48,883	2,44,415	43
1961	55,638	2,78,189	13.8
1971	63,582	2,17,908	14.2
1981	79,633	3,98,168	25.2

(**Source**: Number of households for 39-40 and 1951 is taken from Working Plan of P.N. Khosla (39-40) and A.M. Rai (82-83). Population has been computed @ 5 per household. For the remaining years the population of Tehri Forest Division has been taken as 80% of the population of Tehri District.)

TABLE 8

Population Growth of Tehri Forest District 1961-1981

Year	*Total Population*	*Rural*	*Urban (%age)*	*Decadal increase*
1961	3,47,736	3,40,140	5506	13.60
1971	3,97,385	3,86,862	10523	14.27
1981	4,97,710	4,77,154	20546	25.25

When we look at the fuelwood demand of the seven municipal towns in Tehri district, all of them small in size, we can well understand the kind of pressure being exerted for fuel wood to meet even local urban needs in the district.

TABLE 9

Annual Demand of Firewood in Municipal Areas Estimated Annual Demand (in qtls)

Sl. No.	*Town*	*Fuelwood*	*Charcoal*
1.	Muni Ki Reti	15,000	3,000
2.	Narendranagar	3,000	700
3.	Chamba	2,000	500
4.	Tehri	15,000	3,000
5.	Ghanesali	10,000	300
6.	Kirtinagar	20,000	500
7.	Deoprayag	1,000	300
8.	Grand total	66,000	8,300

(**Source**: A.M. Rai's Working Plan 1982-83.)

The main source of fuel was the oak forests, as this species has both a high calorific value and yields high quality charcoal. The British looked upon oak as a species which stood in the way of conversion of forest areas into commercially exploitable coniferous forests and, therefore, actively encouraged the removal of oak by permitting and encouraging girdling. This was in addition to lopping resorted to by villagers. This policy however, was not followed by the native state and the working plan of Tehri 1939-40 is unique in having an Oak Lopping Working Circle to help and guide village women to lop oak trees in rotation. While the excellent state of oak forests in 1939-40 bears this out, the subsequent depletion would have been caused by ignoring this precept and giving added impetus to the departmental extraction of fuel which therefore came to have a major role to play in decimation of the oak forests, while lopping by villagers also continued, causing inevitable damage to the shrinking pool of oak trees in the forests.

The next table shows the extent of departmental extraction,

which would be in addition to lopping by the villagers to meet their daily requirements.

TABLE 10

Department Extraction of Fuel in Tehri Forest Division 1973-74 - 1980-81 (in qtls)

Year	*Fuelwood*	*Charcoal*
1973-74	29,052	162
1974-75	29,614	752
1975-76	24,972	3,633
1976-77	26,309	221
1977-78	35,411	807
1978-79	32,200	90
1979-80	39,764	153
1980-81	27,193	50

Though the native state of Tehri had a more healthy approach to the oak forests, nevertheless this changed after independence and in the Working Plan of 1950-51 fuel working plan circles were first formed, resulting in 1189 hectares of oak being felled for fuel. Oak is a good coppicer and regenerates if lopped laterally. However, the post-independence rush towards development could hardly have been expected to take care of such small matters. Oak trees were literally beheaded by fuel contractors and left to die a slow, certain death. The Working Plan of 1973-74 prescribes an annual limit of 1,750 trees, but even this contributed to further degradation of oak areas, as noted by A.M. Rai in the working plan of 1982-83. He said,

> It is known that tree cover in banj areas is under stress. Hence, prescribing 1750 trees to be cut per annum was not reasonable. In Chandogi Thathi and other settlement blocks, where biotic pressure was already intense and the oak forest was under heavy lopping stress to meet local demand for fuel, fodder and agricultural implements, the Working Plan of 1973-74 prescribed further felling for the fuel felling circle, as a result of which these forests degraded at rapid rate.

In the chir forests, where oak is intermingled with the dominant species in long tongues down the ravines where they protect the water source, commercial exploitation removed the banj along with the chir. This removal was despite what P.N. Khosla prescribed in the 1939-40 Working Plan. He said,

> The following broad leaved species of economic value, even where in intimate mixture with chir pine, are not to be removed as it is desirable to maintain mixed crops for improving soil factors and for containing fire hazards — banj, oak, sein, sal, sandan, tun, kaphal, walnuts.

Fire in the Forest

Fire has become another major cause of destruction in recent years. One type of fire involves controlled burning of clear felled areas in order to exterminate competitors and achieve a uniform mono-crop of chir. This process is a part of departmental management of chir, a management practice which eliminated associated and adapted species, especially oak, with the result that both chir and oak were adversely affected. Even where burning did not destroy seedlings, it baked the ground, adversely affected soil and moisture conditions making it virtually impossible for the forest flora to regenerate. In addition, much damage was caused by accidental fire and paradoxically (though understandably) this was caused by failure to undertake controlled burning of fire lines (commercial extraction having led to this vicious circle where the remedy was as bad as the disease). Between the years 1973-74 and 1979-80 only 1605 hectares of forests were affected by fire, of which the fire of 1979-80 alone accounted for 1344 hectares. However, between the years 1981-82 and 1986-87, 7930.57 hectares of forest were affected by fire, of which the fire of 1984-85 alone devastated 6897.17 hectares. The period from 1981-82 to 1986-87, ironically enough, saw an enormous increase in funds available for plantation, with no commensurate increase for funds for fire protection. This is symptomatic of the faulty policy being followed in the hill areas of India, especially Garhwal, where lakhs of new trees are planted every year whilst a large number of trees are destroyed by fire for want of protection measures. It is equally significant that whilst fire lines were regularly cut and maintained so long as commercial exploitation of the forests was permitted, with the stoppage of such exploitation in 1981, the maintenance of fire lines also ceased. All the benefit derived from the stoppage of commercial fellings was lost as a result of the increased vulnerability of the forests to fires.

The scant attention paid to fire protection is indicated by the table below, which, gives allocation of funds for fire fighting (it being axiomatic that burning of fire lines and maintenance of coups had earlier been undertaken by the contractors who had bought the coups in public auction).

TABLE 11

Allocation of Funds for Fire Fighting

Year	*Allocation(rupees)*
1982-83	40,000
1983-84	40,000
1984-85	40,000
1985-86	45,000
1986-87	40,000
1987-88	45,000

(**Source:** D.F.O., Tehri)

This positions prevails not only in Garhwal, but throughout the length and breadth of India where forests still survive. Even in Madhya Pradesh, which is India's largest forest state, fire protection has an equally low priority.

Trees Felled by Storms

A phenomenon to be noted in Tehri is that *even after commercial felling stopped, in 1981 the availability of timber, now exclusively obtained from fallen and wind blown trees, remains almost the same as it was when the forests were commercially exploited.* The two tables below give the picture of extraction of timber in the five years prior to the imposition of the ban and the subsequent five years. It can be seen that the quantum of timber extracted from Tehri has remained constant regardless of the fact that formerly it was obtained by selective felling and now it accrues only from fallen trees. Right holders and concessionists continue to fell green trees for house construction to the extent of almost 4,500 cubic metres per annum, since the state policy remains that of harvesting fallen trees for commercial sale though the timber from these trees could have been allotted to the villagers for domestic housebuilding requirements, thereby saving green trees and standing crops from being felled.

TABLE 12

Extraction of Timber 1976-1981

Cubic m (Round)

Year	*Commercial timber*	*Timber for rightholders and concessionists*	*Col.3 as %age of Col.2*
1976-77	8,773	2,220	25
1977-78	8,642	5,445	63
1978-79	8,630	4,114	48
1979-80	4,335	4,995	115
1980-81	3,310	7,120	215
Total	33,689	23,894	
Annual Av.	6,738	4,779	

TABLE 13

Accrual of Timber 1982-1987

Cubic m. (Round)

Year	*Accrual from fallen trees*		*Total*	*Timber for rightholders and conce-ssionists*	*Col.5 % of Col.2*
	Timber	*Fuel*			
1982-83	4,967	2,392	7,359	10,051	137
1983-84	3,495	622	4,027	652	16
1984-85	10,072	286	10,358	3,674	35
1985-86	6,008	788	6,796	3,531	52
1986-87	4,631	580	5,211	3,474	67
G. Total:	29,083	46,680	33,751	21,382	65
Average:	5,816		934	6,750	4,276

One could safely conclude that whilst the area under forest cover has shrunk substantially, the timber being extracted continues at the level which existed even before the famous ban on felling. This is indeed worrisome.

Resin Production and Burnt/Fallen Trees

Both the quantity of resin tapped and income accrued therefrom have also remained constant, which means that large scale tapping of pine trees continues. There is also an apparent co-relationship between the extent of resin tapping and the weakening of trees uprooted in storms. It would seem that fire hazards increase with resin tapping and tapped trees are most vulnerable to fire. Perhaps trees which have been weakened by tapping are also prone to fall down.

TABLE 14

Rangewise Resin Production

(In Quintals)

Range	1984-85	1985-86	1986-87	Total
Bhilangana	2410	2293	2293	6996
Pokhal	864	821	337	2022
Saklana	529	237	89	855
Balganga	339	210	355	904
Total	4142	3561	3074	10777

TABLE 15

Resin Production and Burnt/Fallen Trees

Year	*Range*	*Qntls. produced*	*Resin area burnt (Ha.)*	*Burnt trees (Nos.)*	*Resin wounds (Nos.)*	*Fallen trees (cu.m)*
1984-85	Bhilangana	2410	371	—	—	6356
1985-86	-do-	2293	—	—	—	1517
1986-87	-do-	2293	—	3025	900	845
1984-85	Pokhal	864	816	—	1000	1135
1985-86	-do-	821	10	—	—	793
1986-87	-do-	337	—		—	
1984-85	Saklana	529	1698	2040	—	988
1985-86	do-	237	—	—	—	746
1986-87	-do-	89	1946	—	550	
1984-85	Balganga	339	1723	—	3000	376
1985-86	-do-	210	—	—	—	730
1986-87	-do-	355	—	—	—	—

TABLE 16

Damage by Fire and Trees felled by Storm

Range	*1984-85*		*1985-86*		*1986-87*		*Total*	
	B. T.	*F. T.*	*B.T.*	*F.T.*	*B.T.*	*F.T.*	*B.T.*	*F.T.*
Bhilangana	—	6356	—	1517	3025	845	3025	8718
Pokhal	—	1135	—	793	—	—	—	1928
Saklana	2040	988	—	746	—	500	2040	2284
Balganga	—	376	—	730	—	—	—	1106
Total	2040	8855	—	3786	3025	1395	5065	14036

(B.T. = Burnt Trees, F.T.= Fallen Trees)

The mode of resin extraction has perhaps added to the fire menace. Under the native state, resin used to be collected in earthern pots attached to the trees. As and when the forest floor caught fire these earthern containers either fell off themselves or were easily removed or destroyed. Modern India, however, switched to resin pots made of tin which are nailed firmly to the tree trunks. When the fire spreads, the containers feed the forest with fierce and hard to extinguish flames. The tree literally becomes a sacrificial torch of flame.[2]

The statistics given above show how the forests of Tehri have fallen a victim to so called scientific management. The extinction of oak, sandon, guryal, kaphal, burans and other broadleaved species which co-existed with chir has left the forest floor covered with carpets of pine needles which are prone to catch fire in the piercing heat of May and June (unless fire-lines are cut and controlled burning undertaken). But the latter in turn further tends to exterminate undergrowth of broadleaved species. Clear felled areas where seeding felling was undertaken on a large scale are left with all trees of the same class and age competing simultaneously for limited nutrients.

As explained in chapter four, rootlets of broadleaved trees have beneficial fungi which return nutrients to the soil. The rootlets of chir pine, have an acid sheath on the other hand, which tends to form an acid pan when they join, thus depleting the soil further of all nutrients. This weakens the trees to the extent where lacs of chir trees are then felled by storms.

No doubt the system of having the Van Nigam or forest corporation to extract timber from fallen trees was not the best of ideas. Trees are felled and burnt in remote areas where patrolling is practically non-existent. The infrastructure of protection is woefully inadequate. While the state is officially responsible for extraction of felled trees, on the ground the sawing and removal is undertaken by petty labour, generally from Nepal or petty sub-contractors of the Nigam. Since there is no one to check the extraction, the likelihood of green trees being cut and sawn along with fallen trees cannot be ruled out. Both Sunderlal Bahuguna and Chandi Prasad Bhatt had originally started their peoples movements with a demand to ban, not felling, but non-local, big contractors. The tussle at that time was between the big fish and the little fish, quite reminiscent of the recent struggles of small fishing boats against large mechanised trawlers. Perhaps in those days, local cotractors might have proved themselves worthy sons of the soil. Experience of more recent years especially in the North-East has shown that local contractors can be as great a devastating force as non-local contractors. Infact, greed seems to know no boundaries of race, colour or birth. So long as urban market forces are allowed to operate and govern forest policy, the forest wealth of this planet shall always remain under acute stress.

Management Structure

TABLE 17

Comparison of Area of Territorial Divisions in UP Hills

Sl.No.	*Division*	*Area(ha)*	*Remarks*
1	2	*3*	*4*
1.	Tehri	2,23,964	Includes 33,000 ha of Gangi Reserve (Khatling glacier)
2.	Uttarkashi	4,78,795	Includes 2,31,192 ha snow clad area (Gangotri glacier)
3.	Badrinath	2,14,622	

4.	Tons	1,42,115
5.	Kedarnath	1,39,252
6.	Yamuna	1,19,733
7.	East Almora	1,13,046
8.	West Almora	90,558
9.	Pithoragarh	86,755
10.	E. Dehradun	77,313
11.	Lansdowne	68,818
12.	Garhwal	69,011
13.	Kolagarh	61,094
14.	W. Dehra Dun	47,609
15.	Chakarata	36,168

In the hills of Uttar Pradesh, the Forest Department is organised territorially into Territorial Divisions and functionally into Plantation Divisions and the units of the Forest Development Corporation which are responsible for exploitation of the forest. Tehri District, which is one of the most blighted sections of the hills of western Uttar Pradesh, is covered by a single territorial division to the extent of 53 per cent of the geographical area of the district, with another 12 per cent being administered by the Yamuna Territorial Division and Uttar Kashi Territorial Divisions respectively. 2,875 sq. kms., *i.e.,* 65 per cent of the total area of the district, consists of reserve forests. Thus excluding the approximately 425 sq.kms. of snow covered and blank areas, 55 per cent of the district still consists of reserve forests.

Tehri district consists of three major catchments, the Bhagirathi, the Yamuna and the Bhilangana. This last named river is a tributary of the Bhagirathi. The major part of the district consists of the Bhagirathi and the Bhilangana catchments, with the Yamuna catchment covering only a small portion of the north western corner of the district.

Tehri Forest Division has 12 river catchments, including 4 major catchments.

Some of the main catchments are given below :

TABLE 18

Drainage Systems in Tehri Forest Division

Name of the River System	*Remarks*
Bhagirathi (North)	North of Tehri Bridge
Bhagirathi (South)	South of Tehri Bridge
	The river originates from the Gangotri glacier in Uttarkashi district and is joined by the Bhilangana shortly after old Tehri Bridge.
Bhilangana	Originates from Khatling glacier within Tehri district.
Alaknanda	Runs along southern boundary
Leswalta gad	Sub-catchment of Alaknanda
Balganga	Joins the Bhilangana near Ghamsali
Nailchami	Also joins Bhilangana near Ghamsali
Hyunl	Independent catchment joins Ganga near Shivpuri
Gular	Independent catchment, join Ganga near Shivpuri
Chandnawa Rao	Joins Ganga ahead of Rishikesh
Song R.	Drains Saklana valley, entering Dehra Dun district.

In the hills each river catchment represents a valley enclosed by high slopes and separated physically from the next valley. Trans-valley movement entails crossing of high hills or travelling long distances to points where the valleys converge. This makes supervision extremely difficult. As can be seen from the next table, not only is the terrain difficult and the area large, but in terms of ground staff, 101 forest-guards, (including temporary lopping guard) is all the staff available for patrolling, protection, fire watch and extinction, guarding against theft, as also for plantation and improvement at the beat level. So large is the area, so cut off by rivers and hills, so inadequately staffed that it is beyond human capacity to properly manage the division.

TABLE 19

Area of Ranges in Tehri Forest Division

Range	*Settlement Blocks guards**	*Area of Range(ha) per F.G.(ha)*	*Number of forest*	*Average (app.)area*
Alaknanda	18	37,992	13	3,000
Bhilangana	24	69,110	20	6,300
Balganga	14	40,062	18	3,640
Manknath	41	23,971	12	2,000
Tehri	47	20,678	19	1,800
Saklana	22	22,722	11	2,000
Shivpuri	23	19,428	8	2,400

* including lopping guards.

In the Bhilangana range Gangi Reserve comprising 33,000 ha of priceless forest wealth is under one Forest Guard.

Inadequacy of staff and management resources is further borne out by the next table, which shows that at least four territorial divisions in Kumaon and Garhwal are smaller than the Bhilangana Range of Tehri Territorial Division, which is 691 square kilometres in area:

TABLE 20

Name of the Forest Division	*Area (sq. kms.)*
Pithoragarh	868
Nainital	596
Kalagarh	611
Garhwal	690
Lansdowne	688
W. Dehra Dun	476
Chakrata	362

Quite unlike certain environmental groups which are vociferous in their anti-forest department stand today, as far as back as 1950, Meera Behn was well aware of the constraints under which the forest department had to function and the limitations in execution of their own departmental programmes. Her observations given below are relevant even today.

> The problem of thinning is very neglected, the fire lines are not kept in proper order, the tapping rules are not strictly enforced, and so on and so forth. And one thing is certain, that if best results are to be obtained, the forest department field staff must be considerably increased. At present the Ranger, Foresters, Forest Guards and Fire Guards cannot possibly supervise their range properly. In the Himalaya the Range is always a vast area, and much of it sufficiently difficult of access.
>
> The forest of the Himalaya are the Guardians of the Northern Plains, which, in their turn, are the Granary of India. Surely such guardians deserve the utmost care and attention that the Government can give them.[3]

This wise advice notwithstanding, there are no indications so far, of a reorientation of state policy to strengthen the management structure of the fast depleting reserve forests of Tehri Garhwal and to stem the tide of destruction. Instead, as the second part of this book will show, huge investment in time and money are being made for restoration of the forest in those areas no where now forests remain.

NOTES

1. *Meera Behn [The Hindustan Times, 5 June 1950; Meera Behn: A Centenary Tribute; The INTACH Environmental Series].*
2. Personal Communication to the author from Chandrashekhar Nainwal, DFO Civil Soyam, Tehri in April '87.
3. Meera Behn; *op. cit.*

   All the statistics in this chapter are derived from :

   P.N. Khosla; *The Working Plan for Tehri Forest Division, 1939-40*;

   A.M. Rai; *The Working Plan for Tehri Forest Division, 1982-83*;

   The tabular data also figures in *Afforestation strategies in the Bhagirathi catchment in Tehri Garhwal*, study conducted by Neeru Nanda for National Settlement of Human Rights and Environment, Bhopal.

PART II

RESTORATION

If one talks of the forests of the Garhwal Himalaya, one learns that Garhwal was formerly rich in oak forests, and by looking around the mountains one can see for oneself that slopes which now have nothing but low shrubs on them, were once forests of oak, rhododendron, and other broadleaved species. The picture presented is one of the beautiful oak forests, grass-lands, and fertile fields with plenty of cattle and overflowing pails of milk. The very opposite is the picture of today. It seems clear that the progressive deterioration of the region is due to disappearance of these great forests, and, what is more, the devastating floods caused nowadays in the plains by the waters of the Ganga, appear to be largely due to the same cause. Indeed one is driven to the belief that if nothing is done to grow forests again in the Himalayan catchment area of the Ganga, no amount of money spent on river projects in the Gangetic plain will ever solve the problem.

Meera Behn
[The Hindustan Times, 5 June 1950;
Meera Behn: A Centenary Tribute;
The INTACH Environmental Series].

CHAPTER 7

Social Forestry in the Himalaya

How Common are the Village Commons?

Simultaneous hope and despair are poignantly expressed in the landscape which can be seen from a tea shop below Motana village on the road to Lambgaon in Pratapnagar block of Tehri district. Across the Jalkurgad river, hill slopes are scarred and eroded; landslides and debris slide down innumerable gullies. Not a tree can be seen. It is as if the guts of the mountain have been torn open — not suddenly, but through a slow, inevitable process. Yet a hillslope just adjoining this pitted moonscape is full of greenery and even a few chir trees can be seen peeping above the dense shrubbery and thickets. This plot of about fifty hectares appears to be full of the sap of life, bringing a hesitant smile onto the face of a mountain which otherwise faces desolation and death.

"What happened here?", we asked people at a wayside tea shop. "Did contractors pillage the forests?" They shook their heads. "Abadi bad gayi (population has increased)," they said. "People foraged for what ever they could get".

But why did they spare the adjoining hill slopes? Why is a small forest coming up there? "Oh, that area above Bhenga belongs to Bhenga village, unlike the other hill which belongs to everybody. It was planted by the forest department seven years ago. Chowkidar ki thodi madad mil gayee (appointment of a chowkidar helped)."

Since then, Bhenga villagers have closed the area. Nobody is allowed to cut or lop the shrubs. Slowly, the grass and shrubs are regaining strength and tree tops of a few young chir can be seen amidst the undergrowth. This protected patch acts as a bulwark, preventing soil erosion and ensuring that Bhenga village will not be washed away by landslides and gully slips which mark the adjoining hill where women come to forage for just a few twigs and leaves from both sides of the Jalkurgad river. They come there from distant villages like Motana, perhaps even from Kotgah and Bheluntha. Common land, where usufruct is not limited to a single village or a homogenous group of hamlets, becomes every body's land which ends up as nobody's land now in the hills of Uttar Pradesh. Nobody's land is doomed. Failure rates are high even when the forest department takes it up for plantation. Such terminal soils need treatment in stages, beginning with natural regeneration and protection to facilitate nurture of colonising shrubs and grasses. But in the absence of community controls, the village commons get no rest, since continuous browsing by cattle and foraging for fuel and fodder combine to keep them in a state of degeneration and decay. In degraded sites, where community controls are still extant, biomass density does increase even though planted saplings fail to survive. But in open access lands effective closures cannot take place. Nothing happens except the slow, inevitable slide towards desertification and utter ruin.

While Bhenga villagers have closed the area above their village for ecoprotection, they themselves continue to forage elsewhere for their daily needs of fodder and fuel. So the two hill sides which tower adjacent to each other tell yet another story — it is as if the deserted and plundered slope is telling the other: "You are green because I am plundered. Your gain is my loss. My loss is your gain." And herein lies the present dilemma of community resource management which needs to be planned keeping the entire forest land, of both reserves and civil soyam, in view as an integrated whole. Policies for plantation and restoration have, however, almost exclusively targetted the village common areas in Uttarakhand as an aftermath of the colonial heritage, which still continues to overshadow, not only our policies of exploitation, but also the policies and strategies of restoration and plantation.

Genesis of Civil Soyam in Uttarakhand

Civil soyam protected 'forests' are gram sabha lands which

belong to the revenue department and are formally shown within the revenue boundary of villages or gram sabhas. In this way, they come closer to the definition of village commons at present, since they are loosely regarded as belonging to the people and the gram panchayat has the authority to recommend allotment or lease of civil soyam lands to private individuals or institutions. Legal sanction and control, however, vests with the Deputy Commissioner, the head of the district administration, represented at the village level by the patwari (the village land record keeper). While Tehri revenue records show these as sarkari lands, in British Garhwal civil soyam lands are designated, ironically no doubt, as Protected Forests.

These demarcations, which left narrow strips of land for free use of villagers as a kind of buffer zone between the reserves and village settlements, were an artificial introduction into the native land management system which was destroyed at the time of expropriation of village forests and notification of reserves by the British Government in British Garhwal, and later by the native state of Tehri. In actual fact, these strips of land, just adjoining village settlements, were carved out of a mix of forest, fallow uncultivated waste or benap and winter grazing grounds. They also included common paths, streams, wells etc. There was no question of considering these tracts as community resource banks for biomass production . The Britishers recognised this very well after the fierce agitations and accompanying incendiarism of the forest movement of 1914-19 which ultimately led to the setting up of the Kumaon Grievances Committee and the formal charter of rights accorded to the local people within the notified reserves.[1] It was thus accepted, even in the early 20th century, that civil soyam lands will be unable to fully meet the biomass needs of settlements and because of this constraint the British government, as also the native state of Tehri Garhwal, had to grant widespread rights and concessions within the reserve forests to residents for lopping, grazing and harvesting timber for household construction. A.E. Osmastons Working Plan of North Garhwal 1919 deals with these rights very explicitly.

The indigenous land use patterns had recognised the need to have considerable amounts of fallow land adjacent to the village to permit the practice of traditional agriculture on two and five year rotational cycles.[2] The common lands adjoining villages also used to be kept for winter grazing of cattle returning from chainis or cattle camps situated in the uplands while, for fuel and fodder needs, the villagers used to repair to their native jungles. The forests were divided amongst villages through a system of boundary demarcation about which more evidence would be presented in subsequent chapters. In Garhwal, the existence of traditional boundaries between villages had been documented in the earliest working plans. In 1919 A.E. Osmaston, settlement officer for North Garhwal Forest Division, took into consideration all such traditional boundaries within notified reserve forests while demarcating areas or settlement blocks for exercise of rights and concessions for grazing and lopping. Osmaston noted that even though the villagers could, in principle, exercise these rights and concessions within the entire settlement block, they preferred to confine themselves within their own traditional boundaries as mutually recognised by adjoining villages *within the reserves.*[3]

The Gazetteer of British Garhwal, while explaining the classification of forests into Class I, II and III subsequent to the Forest Grievances Committee 1919-22, adds the noteworthy and intimidating description of Class I forests as those which were waste—'where hardly any forests remain'. The reader who would have followed the accounts of apportionment of grazing areas between local and professional graziers in the Ganges forest division in the late 1880's, in Chapter 4, would not be surprised to learn that these very Class I forests where "hardly any forests remained" were handed over to the district administration to be operated as village forests for the 'benefit of the local people'.[4] Walton himself admits that some of these so-called forests were in a desolate state as early as 1919, after less than 50 years of establishment of the Imperial Forest department. Yet in later years population increase (both human and cattle) would be trotted out as reasons for the grim and barren state of most civil soyam lands.

It was only after the fierce and unrelenting agitation of Kumaon, in 1916-19, that Van Panchayats were created in British Garhwal in the remote and more inaccessible forest tracts stocked with broad leaf species of no commercial value. Today these Panchayats forests mark Chamoli, Almora and Pithoragarh districts as oases in the desert. It is not to be wondered that the villagers treasured their panchayat forests and continued to exploit the reserves for their needs, so that in the post independence period the well stocked panchayat forests have actually flourished at the expense of the reserves. This was only to be expected, in the circumstances, since the entire land use system had been taken over and moulded to the needs of the state while the indigenous management framework had been grossly distorted. But in the absence of adequate land mass, Van Panchayats also could not be worked cyclically and the panchayats evolved working plans based on the principle of rotation and rest only where sufficient area was available.

Multiple Pressures on Civil Soyam Lands

Over a passage of time, the civil soyam came to be increasingly regarded as villagers forests which should cater to the biomass needs of the people, even though this had never been seriously envisaged as their role. Meanwhile, population slowly increased, while extant meagre resources in the village commons were further wiped out. The human population of Tehri district rose from 1.6 lakh at the time of the working plan of 1939-40 to 4.8 lakh in 1981, (an increase of almost 280 per cent over forty years, the decadal growth rate being minimal). Available figures for cattle population show a 90 per cent increase from an estimated 3,20,803 in 1939-40 to 6,25,724 in 1978. On the other hand, A.M. Rai's working plan for Tehri Forest Division of 1981-82 estimated only 10 per cent forest cover for civil soyam lands.

Loss of forest cover and increased biotic stress is not the only adverse factor impinging upon civil soyam lands. As pointed out earlier, these lands included fallow or so called waste lands (often entered as benap in the land records) which actually formed part of the regular two or five year rotational cropping cycle. The convention, (which has been described accurately by Phillip Mason, a British Commissioner of Garhwal) permitted a cultivator to extend his cultivation into the common land adjoining his field and later apply for niyamitikaran or regularisation. Mason further explains that those whose fields did not adjoin the common land were permitted, under the traditional system, to move into remoter areas by crossing a gully or ridge to open up a new plot of land. When land was plentiful there were no conflicts between villagers, but with vast tracts taken out as reserves, the so called commons could not equitably meet up all the demands of an expanding population.[5] However, there was no let up in the British policy of encouraging agricultural expansion — a policy which was also followed by the Indian government.With commons now seen as a scarce resource, it was not surprising that influential or otherwise well placed farmers benefitted more than others by way of extension into these estates. Gradually, most of the common lands got appropriated in this fashion and conflict situations also increased as a result of these fundamental changes in the land management system. Conflict now came to stay as a dominant feature of village life, quite overshadowing ancient traditions of conflict management and cooperation. The patwari replaced the village assembly as a pivotal force, becoming an indispensable part of village politics. The fact that he was introduced by the British Government as an official who was known and expected to live off the land is vouched for by no less a person than the British Commisioner of Garhwal himself.[6]

While the commons bore the brunt of extension of cultivation, which was encouraged by the revenue administration during British rule, this policy was actively continued after independence. Potato and apple cultivation, in particular, were extended at the cost of valuable resources in civil soyam lands, many of which were at that time covered by dense oak forests.[7] In fact, an ambivalent policy, instituted by the British was later perfected after independence. Lip service was paid to the civil soyam as "people's forests". Even today, one arm of the state *viz.*, the forest department talks in terms of civil soyam 'forests' as a buffer zone which is supposed to take the burden off state or reserve forests. On the other hand, the revenue department and district administration, continue to see these common lands as avenues for further agricultural and horticulture expansion and creation of community facilities in rural areas. Civil soyam also has to provide land for the poor by way of housing sites and allotment land for cultivation to the landless labour and marginal farmers.

Though British settlements had followed traditional boundaries while demarcating gram sabha lands, subsequent reorganisation of gram sabhas and political exigencies of Panchayati Raj led to splits and adhoc partitions amongst villages which had traditionally formed a single group. At the grassroots, therefore, a piquant situation has emerged whereby many new gram sabha are vested with dejure boundaries, even as defacto usage rights, within their civil soyam lands, tend to continue, as in the case of reserved forests, along traditional lines.

Increasing urbanisation, coupled with opening up of a vast network of roads, has further led to a situation where the public at large are able to forage for fuelwood, grass and poles along extensive open stretches located by the roadside. The users include outsiders like hoteliers, (catering to increasing tourist and pilgrim traffic) road labourers, contractors and their labour, and the Border roads staff and labour (all of whom have to survive in labour camps set up along the roads they build and maintain). Thus a large part of these lands is subject to multiple consumers over whom neither the dejure gram panchayat, nor the forest department, has any control. As their civil soyam lands get opened up and also appropriated for various schemes and projects, local residents, unable to control the plunder, join in the free for all.

Extent of Civil Soyam Land in Tehri District

Available statistics of the extent of civil soyam lands are not always reliable. Figures were not readily available with the

DFOs or the district administration in Tehri and the study could only elicit guesses which proved to be off the mark. For example, in the district of Tehri three sets of figures available for civil soyam are as under :

(i)	U.P. Forest Statistics	1983	1272 sq.km.
(ii)	Working Plan of Tehri Forest Division	1981-82	545 "
(iii)	Collectorate, Tehri	....	1029 "

However, the figures can be cross-checked by working backwards *i.e.* by deducting the total reserve forest area of 2,87,500 ha plus municipal area of 4,900 ha from the total district area of 4,42,100 ha. The balance 1,49,500 ha would have to be revenue land, out of which little more than half should be cultivated. The District Statistics Handbook of Tehri (1987) reported 73,800 ha, (amounting to 17 per cent of the district's land area) as net cropped area. Hence, the balance area of roughly 75,000 ha could be the approximate area of civil soyam land. Out of this atleast 25 per cent may be deducted for common paths, water channels, rivers and sandy river banks, gullies, landslides and excavated sites, bringing us quite close to A.M. Rai's estimate of 54,500 ha, which may be perhaps taken as the extent of civil soyam land available for serving as a community resource bank in Tehri Garhwal. This is not an inconsiderable land mass, being about 12 per cent of the district area of 4,421 sq km. However, there are no accurate and uptodate statistics regarding the actual availability of these lands for common use of the villagers. The last settlement or bandobast was conducted in Tehri in 1936-37. At a very conservative estimate, about one-third of these lands might now be under actual occupation of individual users. This guess is also supported incidentally by the fact that upto 1987 a total area of 314 sq km had been enclosed by all the plantation divisions put together in the civil soyam and it was common knowledge that saturation point had been almost reached by then. In many parts of Jakholi block it is reported that most of the forest area under civil soyam has been taken over for protection by individual farmers. While this is no doubt good for the ecology, equity is a neccessary and inevitable casualty in this process as the civil soyam area available on ground would be correspondingly reduced.

Problems Faced by Social Forestry in Village Commons

While forest officers cannot be unaware of the ground reality, they seem to have had little say in policy planning which has focussed plantation efforts solely on village common lands. Many have, however, given excellent analyses of the problem and its causes. V.B. Saharia, IFS, Additional Chief Conservator of Forests, Madhya Pradesh has presented a discerning analysis of the problem in a paper in which he conclusively establishes the fact that fallacious assumptions regarding availability of land have been made by the Wasteland Board while laying down ambitious targets for mass plantation. He takes Madhya Pradesh, the single largest state in India, both in terms of land mass and proportion of forest cover, as a case in point, and indicates how an exercise done in connection with the Narbada Sagar project clearly demonstrated that even in the western districts of Madhya Pradesh like Dhar, Jhabua, Devas and Khandwa, which are mostly deficient in forest cover, (with highly arid conditions prevailing in Dhar and Jhabua districts), revenue wastelands fit for compensatory afforestation to the tune of 44,000 hectares required by the project were just not available. Revenue land fit and available for plantation was only to the tune of 1 per cent to 11 per cent of the land as shown available in the district records. Most of the so-called revenue wasteland had been encroached for cultivation or reserved for various other uses. The rest was either too stony or steep or otherwise unfit for planting.[8]

An evaluation report on the social forestry programme published by the Programme Evaluation Division of the Planning Commission in August 1987, found that although one of the main thrust areas of social forestry was to raise block plantations and village forestry on public and community wasteland outside forest areas, the actual performance was disappointing. Out of 256 selected villages only 23 (9 per cent) had raised block plantations on public and community land. In many states the programme was greatly hampered due to acute scarcity of encroachment free public land, while in others the panchayats were either not inclined to handover village commons due to competing demands for these lands or they hardly owned any land on which public plantations could be raised.[9]

In an excellent analysis of the historical process of deforestation, Madhav Gadgil has explained how the policies of the British Government affected non-cultivated lands all over the country, transforming them into open access common property resources. From a system of limited and regulated access to defined villages with specific responsibilities for protection accompanying usage rights, village commons were thrown open to one and all. This is now widely known as the "tragedy of the commons". In the words of Madhava Gadgil, the commons have now become, in the post-colonial period, property that is "everybody's to exploit and nobody's to guard."[10]

The above weaknesses of afforestation programmes sponsored in the heyday of wasteland development are not an uncommon feature of externally sponsored schemes for plantation. In Nepal, the Nepal Australian Forestry project found a close relationship between successful plantations and local motivation. It was seen that people tended to be unmotivated or negatively motivated for plantation either because forest produce was readily accessible already or because of disputes over tenure of plantation land or because the land selected for plantation was already being used for other purposes and consensus for discontinuing that usage was not available.[11] In addition to the fact of dispute over ownership, one could add uncertainty regarding usufruct as a major disincentive for cooperation with forest department plantations in Tehri Garhwal. Thus many so called 'no-mans' lands do fall within legal jurisdiction of various gram sabhas. But since there are too many users from diverse sections of society over whom the dejure owners have no control, they cannot safely hope to enjoy the fruits of enclosure and plantation, even if they are able to temporarily effect closures.

During the Tehri Rajas time the village commons were

respected more than in British Garhwal and it is perhaps due to the original sagacity of the native rulers that the commons still survive even to the extent where atleast 314 sq km of common land could be found and made available for enclosure and plantation. Perhaps this also explains the slightly euphoric vision of village commons in the UP hills, as result of which some models have been prepared comparing upland villages in the latter with their counterparts in the plains. A model prepared by Chambers, Saxena and Shah, contrasting upland/tribal villages with their counterparts in the plains, while not being strictly applicable to all upland villages in U.P., still provides a useful tool for differentiating between villages with better managed commons (and consequently successful plantation programmes) from those where anarchic situations prevail.[12]

Population	50-100 families
Used by	one village
Dependence on wastelands/commons	very high
Authority of village elders	still intact
Market and state penetration	weak
Minimum requirement for regeneration	protection only
Users within a village	all families

To a limited extent, this model can serve to explain why Bhenga could manage to successfully protect and regenerate its adjacent wasteland (with the help of the forest department) while no-mans land continued to degrade right next door. Though this model does not explain all situations where government plantations failed, by and large it can be accepted that ownership, exclusivity, and assured usufruct for a homogenous settlement group seem to emerge as key factors in ultimate success of afforestation efforts of the Forest Department in civil soyam lands. The legal ownership of gram panchayats does not appear to have much relevance. Even while the department is required to enclose areas in consultation with the 'legal' owners, the plantations cannot succeed without the cooperation of actual users who could be either a single contending party or many. It is thus increasingly clear that concepts of social fencing apply to the least problematic villages which actually do not need any external catalytic agents to enthuse them for social fencing. They are already fairly well knit and have an identifiable area which is their own and which they are motivated to protect.

In the no-mans land, social fencing cannot prevail because local management systems are not in existence. Yet clarion calls are made to plant trees and mobilise communities which state policies have long since suffocated. Now, when rigor mortis has almost set in, we expect the corpse to revive and arise Lazarus-like from the graveyard of so called community forests in civil soyam lands. In these tracts only the rule of law can be enforced by a vigilant and strong administration backed by political will. During the course of this study some exceptional cases of successful plantations in no-mans roadside plots were found but in each case the forest department had itself assumed special responsibility for full-time protection. Sanjay Van on the Rishikesh Narendranagar road was one such notable example. In another case, C.S. Nainwal DFO Civil Soyam had successfully grown Italian olives in a small patch just short of Chamba where he had built a hut for a full time chowkidar posted there for its protection.

Biomass produce of the village commons

The gram sabha lands which are left now in Tehri Garhwal contribute towards biomass production which is largely confined to yield of grass for stall fed cattle and small bundles of firewood from coppicer shrubs which have high calorific value and are hardy and drought and fire-resistant.

The widespread shrubs are tungla (*rhus parviflora*) karonda (*carriss* sp.) dhauli (*woodfordia floribunda*) sandan (*eugenia ougenensis*) timla (*ficus roxburghi*). Osmaston mentions many more which were natural associates of chir in the lower zone such as mahua, kathmahua, genthi and others but most of these are now quite scarce. Guryal (*bauhinia varieagata*), mentioned by Osmaston as growing wild, is now hardly to be found except near the homesteads, while aonla (*embilica officinalis*) also mentioned by him is now confined to the Balganga valley only in certain reaches.[13]

Tungla is a hardy and gregarious shrub which can grow into a middle sized tree when well protected. Infact it is nurtured by housewives on the bunds of their fields as an insurance against a rainy day when ill-health may prevent them from foraging further afield for fuelwood. Because tungla seeds profusely and also reproduces through root suckers, it could colonise hill slopes and contribute significantly towards greening the Himalayas in the chir zone. But in the absence of protection and research and recognition of its properties as an invaluable fuelwood, it is being slowly destroyed since it is found most frequently in the civil soyam lands which have little protection and are under tremendous stress and strain induced by multiple users.

Dhauli is another hardy species which is commonly found in the civil soyam, its scarlet sprayed flowers providing a touch of welcome colour in the otherwise drab landscape of these lands in March. In April and May rumex provides generous splashes of deep rose and pale pink, particularly along denuded slopes where rumex is a superb coloniser and soil binder and is seen to be putting up a valiant struggle even against invasive weeds. Karonda is another wonder plant of purely native origin which, though thorny, is still preferred by women as a useful source of fuelwood. It is amazingly hardy and drought resistant (most of these plants, being natural associates of chir, are fire and drought resistant). One karonda plant was observed by the writer in Dang village over a period of time during the drought year of 1987 when a part of the common land was affected by a serious fire. Curiously, this karonda plant slowly turned green and regenerated itself even though there was no rain fall in the intervening period. With the exception of dhauli, however, none of these plants have graced the nurseries of forest department though dhauli, incidentally, is being propagated by the Madhya Pradesh forest department.

The village commons produce grasses some of which are bhabbar, tashila and lunjila (in the chir forest) golda (a hardy survivor and a heavy grass for cattle) musalao (a rich and nutritive grass protected assidiously in Bel village, above Bhaldiyana, as also all along the Kanatal Surkandadevi road). Sailma, an excellent thatching grass, is cultivated and protected in chaks or enclosures by individual farmers as well as by gram sabhas and can be spotted easily all along the Bhilangana valley.

One of the consequences of deforestation and overuse of the gram sabha lands has been the invasion of harmful weeds which are now overtaking the commons. By 1987 large stretches were covered by lantana, eupotorium, euphorbia and agave. Above Tipri, along the Tehri-Jakhanidhar road, dudonea reigns supreme over a stretch of 2-3 kms. Since 1987 new weeds like parthenium and another noxious weed, smelling somewhat like marigold, have also started colonising road side stretches particularly in the vicinity of New Tehri Township and new construction. The worst hit areas are those affected by the Tehri Dam construction and its related activities.

No research was ever done by the department on the produce of the village commons. No doubt the G.B. Pant University and the Central Soil and Water Conservation Institute have conducted research on some aspects of shrubs and grasses. But there has been no meaningful action research on the part of the forest department to make an inventory of the existing wealth and evaluate the potential and its scope for further augmentation.The value of these shrubs was established during the course of this study when it was found that the majority of women respondents were dependent on shrubs rather than timber trees for fuelwood. Thus the products of the commons are still sustaining villagers even in their attenuated state and there is alot of competition amongst various user groups for extracting the available biomass.

The Reserves as Biomass Banks for People

Had government been in right earnest in its desire to plant trees these would have been best planted in reserves which have been the real biomass banks of the people even after their expropriation by the state. The forest reserves are still located at a relative disance from the diminishing influence of roads and urbanisation and the soil and climatic conditions are such as to be more conducive for efficient and quick afforestation. Informal village boundaries are still recognised in the reserves and while these are formally recorded in the settlement of rights in former British Garhwal, it will be demonstratively established, as we go along, that these boundaries also exist and are jealously guarded and respected in the reserves of Tehri Garhwal. Hence villagers continue to identify deeply with the reserves and this feeling has been enhanced after the chipko successes and the ban on felling of green trees above 1000 metres. Most importantly, the reserves are not subject to pressure of multiple users to the same extent as the so-called commons. It is significant that, in Tehri Garhwal, during the drought year of 1987, there were no quarrels over grass collection by the roadsides in the no-mans land. Women used to come from remote villages traversing distances of 50 km to 60 km to collect grass from village commons situated all along the Chamba-Narendranagar and Deoprayag-Rishikesh highways. However, a great hue and cry was raised whenever the same women entered remote oak forests above village settlements in the reserves where traditional boundaries are still being maintained (though over eight decades have passed since rights to these were legally terminated and replaced by concessions). This happened in Ghamsali area where local women objected strongly to these new inroads into what they regarded as their preserves within the state forests.

Thus, we find a somewhat piquant situation arising in the field where reserve forests, rather than village commons, now carry the greatest extant usages and customary boundaries between villages and consist of a series of resource pools which the latter identify with, not simply for exploitation, but also for protection and management, even as dejure control remains with the forest department. *It is a strange paradox, therefore, that reserve forests sometimes belong more to the villagers than the civil soyam lands or village commons.* Had this fact been recognized earlier, the direction of planning for afforestation could have been changed to give it a more meaningful thrust, with more investments being directed towards reserve forests which are actually feeding the people more than the village commons, but are quite starved of funds. Social forestry should have targetted the reserves as well as civil lands by being flexible and responsive to peoples needs and site conditions. Instead it was governed by ideologies which regard the so-called commons as biomass banks for the people while preserving reserve forests for the state. But to view the reserves as a source of biomass for the local residents would have required clarity of perception and a grasp of the historical context in which the commons were carved out and the reserves expropriated. It would have called for rewriting not only the National Forest policy but history itself. The forest department and government would have to change their perceptions of history, their understanding of the past and their expectations from the present. In the decade of the seventies, when the social forestry programmes were first drawn up, this would have been difficult, even though, as a pointer and beacon showing the way, the Arabari experiment was started in this very decade in West Bengal.

NOTES

1. G.B. Pant, *The Forest Problem of Kumaon* (ed) Ajay Rawat reprint Gyanoday Prakashan, Nainital.
2. H.G. Walton. *Gazetteer of Garhwal Himalaya* Reprint, Nataraj, Dehra Dun 1989 pp.34-35 gives a description of crop rotation and definition of wasteland (p.103) as "land which was previously measured but was found *out of cultivation* at the last settlement was recorded as the property of the government." This policy directly led to extinction of rotational cycles, land was not allowed to rest, the decline of agricultural productivity started.
3. A.E. Osmaston. *Working Plan of North Garhwal 1919*, p. 8.
4. H.G. Walton. *op cit* p. 17. "Class I comprised forests which had practically disappeared and where not worth saving... these open forests were made over to the Pradhans and Patwaris".
5. Philip Mason. *A Shaft of Sunlight*, Vikas Publishing House, printed in Great Britain, 1978. p. 132.
6. *Ibid.*, p. 127..in his memoirs of his tenure as Commissioner,British Garhwal, Mason, suggests that "for a reasonably honest patwari the

job was worth atleast two hundred rupees, more than twelve times the pay...it was a very loose administration traditionally everyone gave him a present of grain according to the size of holding..an 'honest' patwari would accept presents in kind and take what were really fees for appearances in court." One shudders to think what the dishonest patwaris would be upto, even as per the Commissioners assessment and standards of behaviour expected (or rather not expected) from this class of government servant (an institution which was entirely the creation of British rule in Garhwal).

7. For socio-economic impact of cash crop cultivation in Tehri Garhwal see Manjari Mehta, *Cash Crops and the changing context of womens work and status: a case study from Tehri Garhwal.* (ICIMOD discussion paper series Jan 93).
8. V.B. Saharia IFS Additional Chief Conservator, Madhya Pradesh unpublished paper.
9. Programme Evaluation Organisation, Planning Commission, *Evaluation Report on Social Forestry Programme.* New Delhi 1987.
10. Madhava Gadgil,'Deforestation:Problems and Prospects' in *History of Forestry in India*; edited Ajay Rawat, Indus Publications, New Delhi 1991.
    Anil Agarwal and Sunita Narain, *Towards Green Villages*, Centre for Science and Environment, 1989, p.12 'the alienated commons.'
11. R.J. Fisher, Hukum Bahadur Singh, Deepak Raj Pandey and Helmut Lang, *The Management of Forest Resources in Rural Development: A case study of Sindhu Palchok and Kabhre Palanchok district of Nepal,* Nepal Australia Forestry Project, ICIMOD Joint case study (unpublished) hereafter referred as *Fisher et al.*
12. Robert Chambers, N.C. Saxena, Tushaar Shah. *To the Hands of the Poor : Water and Trees,* Oxford and IBH Publishing Company Pvt. Ltd., 1989, p.162.
13. A.E. Osmaston, *Working Plan* p.18.

CHAPTER 8

Survival Rates of Government Plantations

In the early part of the decade of the seventies, when policy makers woke up to the extent of degradation of village commons, as also to the acute shortage of fuel and fodder being faced by the common man, huge investments in afforestation were directed towards restoration of people's village woodlots in civil soyam estates. This decade saw the inception of the Soil Watch programme in the U.P. Himalayas and setting up of Civil Soyam Plantation Divisions, directly under the U.P. forest department, in each district of Uttarakhand. Investment further accelerated rapidly in the 1980's, which have seen the heydey of wasteland development and investment in social forestry. The Congress Party included restoration of wastelands and plantation of lacs of trees in the party's twenty point programme. The State Plan sector in the VI Plan targetted 42,500 hectares of revenue civil soyam land in villages for development of pastures and afforestation in the U.P. hills. The total investment for the plan period was Rs.26.93 crores.[1]

In the same decade, a World Bank aided scheme for watershed management was taken up with a planned investment of Rs.6.3 crores in the eroded areas of the Nayar and Paunar catchments falling in Pauri Garhwal. The European Economic Community funded the South Bhagirathi five year project in Tehri. A Rs.10.0 crores compensatory afforestation project for the catchment area of the Tehri dam started work in 1982-83. This scheme, when operationalised, also targetted non-forest lands. Hence the pace of investment for afforestation in revenue commons was very high.

It is interesting that while thirty years of planning had barely achieved a plantation coverage of 622 sq km in the reserves, the state government set itself a target of 425 sq km for village commons in a single plan period.[2]

During the period 1975-76 to 1987-88, the oldest and largest Civil Soyam Plantation Division covered 11,800 ha at a cumulative cost of approximately Rs. 2.83 crores (representing plantation costs only). Total departmental expenditure (including salary bills of chowkidars) amounted to Rs. 4.18 crores, with an annual average expenditure of roughly 40 lacs. Plantation costs per hectare rose gradually from about Rs. 2,000 per hectare at the inception of the division to about Rs. 4,000 per hectare in 1987-88. In Tehri, between 1975 (when the first Civil Soyam Plantation Division was set up) and 1988 (when the study was conducted), about 31,400 ha of civil soyam areas had been officially enclosed for plantation by the forest department under various social forestry programmes. An abstract of statistics covering most of the period under study is given below:

TABLE 1

Plantation and Expenditure in Civil Soyam Scheme 1976-87 by various Divisions

Division	*Period Covered*	*Area(Ha) Planted*	*Trees(Lacs)*	*Expenditure Rs.(Lacs)*
Civil Soyam	76-87	11865	93.63	283.78
Soil Conservation	83-87	2444	47.01	71.71
Tehri Dam I	83-88	4042	42.20	107.06
Tehri Dam II	83-87	3560	42.21	105.58
S. Bhagirathi	85-87	3068.	—	—
Total	76-87	24979	225.05	568.13

Enclosure of 314 sq km out of 545 sq km of civil soyam was no mean achievement. However, fairly extensive data collected during the course of this study indicates that survival rates of plantations were in no case commensurate with the accelerated pace of investment. The target of afforesting wastelands was not achieved, nor did the scheme attain its declared objective of meeting biomass needs of the people by planting fast growing fuel and fodder species.

Survival Rates of Civil Soyam Plantations

Several sources were tapped for gathering data pertaining to survival rates. One set of information was gathered through field staff who were temporarily employed to undertake a cent per cent enumeration of 40 selected plantations in civil soyam lands covering an area of 508 hectares. Species wise survival rates were worked out by matching the data with the original plantation records. However, the figures so obtained were not representative of all plantations, since site selection was undertaken keeping in view the need to include different types of climatic zones and different plantation divisions in the sample. Selection was also biased by constraints of time and resources and by the requirements of the study which sought to pinpoint factors leading to success or failure. This led to a conscious decision to include representative polar types in the sample which thus reflected results of some unusually successful plantations.

Data regarding 31 sites is given below :

TABLE 2

Enumeration Results of Sample Plantations

Division	*Block & Microclimate*	*No. of sites*	*Year*	*Area (ha.)*	*Trees planted*	*Trees survived*	*%age*
TD I	Bhilangana	11	84-85	6	4,975	2,975	60
	(High rainfall		85-86	58	77,300	52,768	68
	area with better		86-87	41	67,600	60,507	90
	site conditions)			105	1,49,875	1,16,250	78
TD I	Jakhanidhar (Dharkot)	8	84-85	33	47,600	28,516	60
	(Rainshadow adverse		85-86	70	91,950	41,370	45
	site conditions)		86-87	5	7,000	3,242	46
				108	1,46,550	73,128	50
TDII	Thauldhar	9	84-85	33	51,520	22,125	43
TD II	Pratapnagar		85-86	23	37,200	11,350	30

SC	Jalkurgad (Rain-shadow areas with adverse site conditions)		86-87	10	16,000	7,000	44
				66	1,04,720	40,475	39
CS	Narendranagar (High rainfall area & favourable site)	3	84-85	13	12,330	6,335	51
			85-86	20	32,000	23,397	72
				33	44,330	29,397	61
Grand Total		31		312	4,45,475	2,59,250	58

It can be seen from the above table that survival rates of two to three year old plantations ranged from 60 per cent to 43 per cent. However, in certain plantations in Bhaldiyana and Dang (not included in the above table) the survival rates were as low as 5 per cent.

While the enumerations yielded valuable facts that give a realistic picture of the ground situation, for reasons stated above, it was not possible to draw macro-level conclusions from this limited survey. Hence, Deputy Conservators of Forest of various plantation divisions were requested to provide their department figures gathered through periodic enumerations undertaken by the existing machinery.

Survival rates of 70 plantations of the civil soyam divison (as on March, 1986) are given below:

TABLE 3

Survival Rates : Civil Soyam Division
(As on 31/3/1986)

Year	*Total No. of plantations*	*More than 50% survival*	*Less than 30% survival*
1981-82	10	2	8
1982-83	22	Nil	22
1983-84	11	Nil	8
1984-85	16	8	8
1985-86	11	8	3
Total	70	18	47

TABLE 4

Survival Rates : Soil Conservation Division
(As on 31/3/1986)

Year	*No. of trees planted**	*Average survival %*
1980-81	1.11	25%
1981-82	2.98	38%
1982-83	2.11	43%
1983-84	8.58	38%
1984-85	7.81	46%
1985-86	8.97	80%
Total	31.56	

*In lacs.

From the above it is clear that the older plantations indicate a tendency to die out while the more recent plantations (two years and less) show higher survival rates.

A similar trend was found in statistics collated from the department assessment carried out by the Soil Conservation division in March,1986.

While the average survival rate of six years was 43 per cent for the plantations enumerated in Table 3, it is clear from the cumulative figures given in Tables 3 and 4 that the older plantations were showing a marked tendency to die out. Site surveys revealed that surviving plants were mostly weak and straggly and were not such as to hold out much hope for the future of the plantation programme as a whole.

Field observations showed that the survival rates were in fact even lower since wide divergence surfaced in the figures of survival as reported by the department and the actual survival as found at site by the retired forest guard appointed by the writer to enumerate selected plantations. Figures with respect to three plantations under the Civil Soyam Division are given in Table 5. The difference in surviving plants is quite considerable and, in the circumstances, disturbing. It is likely that, given the mass plantation targets and hectic activity involved, quality control by senior officers remained weak in some of the large and unwieldy divisions like the Civil Soyam Plantation Division.

TABLE 5

Year	*Name*	*Trees planted*	*Survival rates Dept.figures*	*Survival rates Field enumeration*
1985-86	Piplethdor	10,912	7,500	622
1983-84	Banali III	36,788	12,984	2,028
1983-84	Pipleth II	23,800	12,616	1,721

Survival rates of plantations in the Reserves

A superficial comparison of the survival rates in the plantation divisions and the reserve forests, might lead us to believe that setting up of dedicated plantation divisions has achieved better results. The table below would give this impression:

TABLE 6

Rangewise Survival Rates of Reserved Forest Plantations 1952-80 (As in March, 1982) Tehri Territorial Division

Range	*Total area Planted*	*Total Plants Planted*	*Plants survived*	*Survival Percentage*
Shivpuri	638	5,33,262	64,673	12
Tehri	534	4,95,260	93,928	19
Saklana	329	3,32,483	95,944	29
Maniknath	390	3,68,535	52,802	14
Alaknanda	375	2,97,073	47,902	16
Balganga	565	6,98,580	46,569	7
Bhilangana	409	3,95,829	21,083	5

(**Source** : Working Plan of Tehri Forest Division. 1982-83—A.M. Rai)

TABLE 7

Area Planted under Plan and Non-Plan in U.P. Hill Reserves

Area in ha

Schemes	*Upto 3rd Plan*	*Three Annual Plans*	*IVth Plan*	*Vth Plan*	*Annual Plan 1979-80*	*1980-81 and 1981-82*
1	*2*	*3*	*4*	*5*	*6*	*7*
Plantation of species of Eco.& Ind. Importance	4,990	3,998	8,542	4,317	2,274	5,917

Rehabilitation of degraded forests	2,920	-	-	-	-	-
Under Soil Conservation	6,309	-	-	-	-	-
Area Planted under Non-Plan	19,154	3,775	-	-	-	-
Total	33,373	7,773	8,542	4,317	2,274	5,917

(*Source*: Forest Statistics of U.P. 1981-82.)

The figures given above need to be carefully analysed from various angles if we are to correctly absorb the message which they convey. They are further interesting because the enumerations span the entire post-independence period giving us a birds eyeview of the performance of independent India with regard to care and protection of the reserve forests.

It is evident that in thirty years since independence the government planted only 35 sq km of area in the reserves of Tehri district. This is a mere pittance when compared both to the pace of exploitation of the reserves, as also the huge investments made in the civil soyam. Infact it is obvious from these figures that the government has never given serious thought to replenishment of the reserves of Tehri district which cover an area of over 2875 sq km. The fact that this neglect was typical of the treatment meted out to the U.P. hills is borne out by figures given in Table 7.

The survival rates documented by A.M. Rai do indicate that some genuine woods (no matter how small) were coming up, since the enumerated plantations date back to the 1950s and 1960s and survival rates of even 20% are significant when they pertain to sites which are over twenty years old. This observation is also borne out by eye surveys and extensive tours conducted by the writer in Tehri district over the past ten years since inception of this study. One is constrained to record that the plantation divisions have next to no single wooded area to show for all the effort and money poured in since the commencement of the programme twenty years ago. On the other hand, the reserves carry the only extant woods which have been planted and reared by the department, though it must be confessed that most of the large ones were raised during the Rajas time, while very few date from the post-independence period. The deodar plantations of Thauldhar (later popularised as the Dhanolti forests) are mentioned as receiving sufficient care in the Working Plan of 1939. The beautiful blue pine and deodar forests of Pratapnagar are another outstanding example of departmental plantations raised by the native state and, significantly, these are seen to have nurtured and co-existed with the broadleaved forests. Osmaston mentions the first surai plantations taken up around Nainital in British Garhwal.[3] But he also mentions the rather disappointing fact that experiments for regeneration of local species did not meet with much success. However, these plantations have now turned into completely naturalised forests and are seen to be coexisting peacefully with native broadleaved species. Perhaps the rootstock of native species simply regenerated itself in the protected areas where plantation of surai, deodar and blue pine was attempted. While the survival of conifers proved higher in the plantations in the reserves, and therefore the departmental bias towards plantation of these species was to that extent understandable, walnut and oak were also seen to perform well. The figures given below bear this out to a certain extent.

TABLE 8

Species wise Survival Trends in Reserve Forest Plantations where overall Survival was more than 20 per cent (Period 1951-March, 1980)

Sl. No.	*Species*	*Trees Survived*
1.	Deodar	75,001
2.	Walnut	10,421
3.	Oak	9,949
4.	Chir	9,476
5.	Blue Pine	8,651
6.	Bamboo	4,936
7.	Kharik	4,135
8.	Miscellaneous	12,175

(*Note* : Due to insufficient matching data regarding species planted and

survived, total number of surviving plants of main species have been given.
Source : A.M. Rai Working Plan of Tehri Division 1982-83).

Species selection and species wise survival rates

Ongoing evaluation could have helped the department to take corrective measures in atleast some segments, even if systemic changes could not be effected. One such area was that of choice of species which could have been remedied even within the given parameters of the programme.

A detailed analysis of the results of cent per cent enumeration revealed the startling fact that the department was blindly planting certain species in overwhelming numbers even though these species had the lowest survival rates. Species with higher survival rates, however, were not being so widely planted. Figures in this regard are given below :

TABLE 9

Species Wise Plantation in 508 Hectare Enumerated

Species	*Total planted*	*Survival %age*	*Rank by %age of plantation*	*Rank by %age of survival*
Dainkan	1,63,889	27	1	6
Bhimal	49,833	4	2	5
Kharik	30,219	23	4	7
Kachnar	16,620	41	6	4
Walnut	40,445	65	3	2
Chir	27,599	66	5	1
Robinia P.	7,353	49	7	3

While the above figures are based upon actual enumerations undertaken during the course of the study, similiar species wise survival trends show up in the departmental enumeration of sample plantations. The data is given below :

TABLE 10

Species wise Survival in Department Enumerations

Species	*Total planted*	*Total survived*	*Survival Rate*
Bhimal	33,862	15,290	45.15
Dainkan	1,11,362	40,609	36.46
Kharik	20,534	6,330	30.83
Kachnar	11,293	6,175	54.68
Walnut	27,482	24,116	87.75
Chir	18,753	16,763	89.64
Robinia	10,206	6,699	65.64
Total	2,33,492	1,15,982	49.67

It is obvious that over 70% of the trees being favoured by the department had less than 45% chance of survival.

Dainkan (persian lilac) and bhimal (grevia optiva) suddenly began to be planted during the great spurt of spending on social forestry which started in the 1980's in the UP Himalayas. These were not only fodder trees acceptable to the villagers, they were also easy to propogate, the seed being widely available and germination rates high. Historically and ecologically, bhimal, dainkan and kharik (celtis australis) are seldom found growing in forests. Instead they generally thrive along bunds of fields in the village setting where they can derive maximum benefit from soil, moisture and manure. In local parlance they are commonly called 'Awaz Pasand' trees, *viz.* trees which thrive better near human settlements where the sound of human voices carries. Waltons District Gazetteer of Garhwal noted as early as 1910 that the "bhyunl, an extremely useful tree, grows in the valleys and lower hill slopes and is *carefully protected by the cultivator*, for its leaves afford a very excellent fodder for cattle while its shoots yield a fibre from which are made the ropes used by the inhabitants of the lower hills. *It is usually grown on the walls between fields*".[4]

A household survey conducted in Bel and Dang villages of Tehri district revealed 545 bhimal and 142 dainkan trees being nurtured by 32 families which together had a total land holding of about 646 nalis (about 32 acres). This gives an average of twenty fodder trees per acre on private farm land

and bears out the common saying regarding the fact that these are "awaz pasand" trees.[5]

TABLE 11

Type of Trees owned by Sample Respondents (Villagewise)

Village	*Chir*	*Bhimal*	*Tungla*	*Total*	*Fruit*	*Total*
Bel	183	351	981	1515	116	1631
Dang	593	687	337	1617	151	1768
Ghuttu	29	160	25	214	49	263
Patagli	113	450	192	755	192	947
Total	918	1648	1535	4101	508	4609

TABLE 11A

Average No. Trees per Household/per Nali (Villagewise)

Village	*Av. No. of fruit trees per household*	*Av. No. of other trees per household*	*Av.No. of fruit trees per nali*	*Av.No. of all trees*
Bel	5.8	76	0.37	5.16
Dang	4.7	51	0.23	2.70
Ghuttu	1.4	6	0.10	0.53
Patagli	5.3	21	0.31	1.54

Average number of fruit trees per sample household = 4.12
Average number of other trees per sample household = 33.34
Note : 20 Nalis is equal to approximately 1 acre.

The figures given above show that *the trees which the department sought to plant in huge numbers, despite the risk of high mortality rates, were infact being nurtured very easily by villagers at no expense.*

Perhaps the Chipko movement and the insistence of chipko activists for planting broadleaved fodder species had also influenced these choices. In Tehri district, chipko activists occasionally visited nurseries to uproot chir seedlings and while their antipathy towards chir certainly had a basis in ecological and historical facts, they too were unable to provide any concrete guidance to the department with regard to alternate strategies.

Elsewhere in the country, during this whole period, the forest department had to work under compulsion of large-scale plantations targets, to fulfil which mass scale plantations of limited number of species were usually taken up. In the plains the programme was being hijacked by eucalyptus or poplars.[6] In other areas it was dominated by subabul or massive plantation of prosopis (gainda babul). In certain parts of the south it was largely confined to plantation of casurina. Hence the forest department in the U.P. hills can hardly be blamed for mass planting of a few selected species which were easy to propogate.

Ironically, in the plains and western ghats, the department selected commercial species which survived well but did not meet fodder needs at all while fuel needs were met only incidentally. In the U.P. hills, on the other hand, fodder species were selected but were planted on a mass scale in degraded areas where they did not have a chance of growing into useful forests.

In any case, the only forest belts of fodder and fuel trees to be found growing naturally in the Himalayas are the oak forest and unhappily and predictably, plantation of oak trees did not figure in the programme even though banj (oak) was a widely preferred species by all villagers.

In the sphere of social forestry therefore the department seemed to be facing a catch 22 situation, which has been very well described by Anil Agarwal and Sunita Narain.[7] This basic dilemma also emerges from figures collected by the Indian Institute of Public Opinion in a Tree Survival Survey of five states covering the same period as the study *viz.* 1983-84 to 1987-88.[8]

Survival Rates of Social Forestry Plantations in U.P. Plain Area

In the summary report on the tree plantation programme in Uttar Pradesh, the Institute found an average survival of approximately 59 per cent with the oldest plantations also showing a lower trend of approximately 50 per cent which increased to 70 per cent for plantations of 87-88. The average survival rates were about 53 per cent in village woodlots and strip plantations and 64 per cent in the forest department land. Since hardly any plantation was done in forest department land in U.P. hills, the village woodlot average of 50 per cent in the plains may be taken as the benchmark for comparison. It is higher than the average survival rates in the hills as shown in Table 2 and Table 3. However, the reason for higher survival rates in the plains becomes evident when the species wise break-up is examined. The study made by the Institute of Public Opinion found that eucalyptus and sheesham together accounted for over 47 per cent of all seedlings planted and both had high survival rates of 71 per cent and 64 per cent. Teak and bamboo were planted in fairly large numbers in forest department lands where they recorded survival rates of over 78 per cent and 63 per cent respectively. The figures given below indicate the survival trend as also the plantation index for selected species.

TABLE 13

Village Woodlots in U.P. Plains

Species	*Rank by % of plantations*	*Rank by % of survival*
Miscellaneous	1 (34.08)	4 (45.61)
Sheesham	2 (25.45)	1 (64.72)
Eucalyptus	3 (21.26)	3 (58.78)
Babul	4 (7.20)	2 (58.78)
Bamboo	5 (6.58)	2 (58.78)

TABLE 13 A

Strip Plantation in U.P. Plains

Species	*Rank by % of plantations*	*Rank by % of survival*
Eucalyptus	1 (46.66)	4 (41.36)
Miscellaneous	2 (26.02)	2 (65.30)
Sheesham	3 (12.75)	3 (60.44)

TABLE 13 B

Plantations in Forest Department Lands in U.P. Plains

Species	*Rank by % of plantations*	*Rank by % of survival*
Miscellaneous	1 (25.90)	6 (59.72)
Sheesham	2 (20.71)	3 (64.13)
Eucalyptus	3 (13.74)	7 (58.56)
Bamboo	4 (10.43)	4 (63.39)
Arjun	5 (8.68)	2 (73.80)
Teak	6 (7.83)	1 (78.20)
Babul	7 (5.58)	5 (60.35)

It is clear from a comparison of the figures given in above tables that the plantation programme of the U.P. hills alone was heading towards disaster, while the department in the plains was making fairly correct choice of species by way of survival rates, though even in the plains the choice of species could have been bettered *e.g.* eucalyptus ranked seventh by way of survival but accounted for one third of the trees planted. But the U.P. hills were under greater local pressure to regenerate forests for fodder and fuel, and it is not surprising that the department took the easiest way out, instead of exploring alternative models to pinpoint appropriate strategies for meeting fodder and fuel needs of local people.

Reasons for failure as perceived by the people

While conducting village studies an attempt was made to ascertain people's views on the afforestation programme taken up by the forest department. Here two sharply divergent views emerged between the Bel-Dang group on the one hand and the Patagli-Ghuttu group on the other. In the first group of villages the survival rate of plantations was less than 10 per cent, whereas in the latter it ranged as high as 60 to 80 per cent. Obviously, where the plantations were not successful, the people also felt that they had derived little benefit from the programme. The table below illustrates this point.

TABLE 14

Benefits of Afforestation (Villagewise) as Reported by Sample Respondents

Village	*No benefit*	*Availability of grass*	*Increase in forest area*	*Both (Col 3 &4)*	*Total respondents*
Bel	18 (90.0)	1 (5.0)	—	1 (5.0)	20
Dang	24 (75.00)	7 (21.88)	1 (3.12)	—	32
Ghuttu	4 (11.43)	20 (57.14)	—	11 (31.43)	35
Patagli	—	3 (8.33)	27 (75.00)	6 (16.67)	36
Grand	46	31	28	18	123
Total	(37.40)	(25.20)	(22.76)	(14.63)	(100)

Curiously enough, the villagers assessment exactly coincides with the figures collected and presented here as also with the writer's observations based on extensive field coverage. Out of 123 respondents, only 46 actually reported increase in forest cover and when 36 respondents of Patagli are excluded the proportion is reduced considerably *i.e.* out of 87 respondents only 13 [about 12%] felt that they had benefitted by way of tree plantation. However, 39 or a little less than half did feel that the programme had resulted in increased availability of grass. *Hence the people seem to have got neither trees nor money out of the plantation programme in the commons. But by and large they did get grass. It is a sobering thought that this benefit could have been derived at one tenth of the cost to the public exchequer.* Had policy makers and senior bureaucrats been clear about the history of the village commons and the kind of multiple pressures operating therein, perhaps a different and less cost intensive strategy could have been adopted. Village commons could have been simply enclosed and protected, at minimal expense, with the aim of obtaining increased yield of grass and native shrubs having high calorific value.

The next table shows the villager's views on why plantations have failed.

TABLE 15

Main Reasons for Failure to Department Plantations as Reported by the Sample Respondents (Villagewise)

Village	*Not reported*	*Adverse Climate*	*Improper Plantation*	*Inadequate After care*	*Lack of Public cooperation*	*Total*
Bel	4	5	8	2	1	20
	(20.00)	(25.00)	(40.00)	(10.00)	(5.00)	
Dang	13	3	2	11	3	32
	(40.63)	(9.37)	(6.25)	(34.37)	(9.38)	
Ghuttu	1	12	1	6	15	35
	(2.86)	(34.29)	(2.86)	(17.14)	(42.85)	
Patagli	-	30	1	4	1	36
	(83.33)	(2.78)	(11.11)	(2.78)	—	—
Total	18	50	12	23	20	123
	(14.63)	(40.65)	(9.76)	(18.70)	(16.26)	

In Bel, improper plantation technique was considered the main cause of failure, whereas in Dang it was inadequate after-care and protection. This is also reflective of the position of the two villages. Bel is relatively inaccessible, and would conform to the model developed by Chambers, Shah and Saxena regarding upland villages where plantation programmes could be successful because of better local management of their village commons. However, success would be possible only if the proper strategies of plantation and protection are adopted. (The next chapter will show how defective plantation strategies worked against the programme, even where the commons were jointly protected and villagers were anxious to raise plantations.)

Dang, (where civil soyam lands adjoin the roadside and are under heavy pressure of multiple users), plantations were almost a complete failure. Infact the survival rate was barely 5%. Dang could be placed at the opposite pole of the model of upland village and the apathy and alienation of the villagers is aptly illustrated by the fact that 13 out of 32 respondents failed to express any views at all regarding the afforestation problem, while 34% highlighted inadequate after-care and generally felt that the department was to be blamed for the dismal state of the programme.

In the Patagli and Ghuttu group of villages the gram sabhas were more deeply involved with the management of common land. For example, in Patagli, the gram sabha had taken intiative to enclose three or four plots in civil soyam and one plot in the adjacent reserve forest. Protection provided was extremely effective. When an attempt was made by one of the families to encroach common land, the gram sabha handed it over to the forest department for plantation. In village Bhergaon the gram sabha demonstrated equal interest in protecting common land and making it available to the forest department for plantation. This interest in afforestation was quite typical of the Ghuttu valley, but the civil soyam area was plagued by revenue disputes between villages since extensive oak forests still survived in the civil soyam and invariably gave rise to ownership disputes, sometimes followed by protracted litigation. Hence, even though survival rates of plantations in the Ghuttu valley were much higher than in Bel and Dang, and even though the villagers were motivated for plantation, the inter-village disputes complicated matters and sometimes actively obstructed afforestation efforts of the department.

The villager's views about the forest department are definitely coloured by performance. As already stated, in Bel and Dang the villagers did not view the department with favour because there was no tangible increase in availability of forest produce. Ghuttu area on the other hand, had witnessed a dramatic increase in grass availability and high survival of planted saplings and Patagli saw rapid growth of shrubs and planted trees. The plantation division covering Patagli and Ghuttu area was a compact, though large, division which had been headed since inception by young and hardworking All India Forest Officers, Shri Ashwini Kumar, IFS followed by S.T.S. Lepcha, IFS. The Range Officers D.P. Bhatt and B.D. Singh were also very dedicated and supervision of field staff was close. This had the neccessary impact on people, as a result of which only 5% and 7% of the respondents of Patagli and Ghuttu blamed the department for failed plantations. *This is an extremely encouraging sign which demonstrates an abiding truth, that, given sincere workers to implement even a flawed policy, the common man can and does respond positively to the government machinery and government programmes.*

As already mentioned, the Bel-Dang group and the Ghuttu-Patagli group fall into two polar types of settlements which have come to categorise Uttarakhand. The latter fall in the category of villages which are homogenous in character with relatively strong community bonding amongst themselves and exclusive usufruct rights over their village commons, these rights being jealously guarded though they are often subject to fierce inter-village disputes. Still they are distinct from the common lands of Dang or the hillside adjoining Bhenga, where all and sundry forage freely, pulling out valuable shrubs from the roots. Nobody exercises control over these lands, the dejure owners being incapable of doing so, while the government also perhaps lacks the vision or the will. Hence, it would be safe to conclude (if generalisation have to be made) that for plantations to succeed three essential requirements have to be fulfilled-there should be an appropriate strategy of plantation and protection coupled with a cohesive village community having exclusive usage and effective control of its village commons, along with alternative areas to shift biotic stress consequent upon enclosures. The next chapter recounts the story of tours undertaken by the writer, mostly in remote villages, away from the roadside, in search of failed as well as successful plantations, in order to get an idea of how plantation strategies actually worked on the ground.

NOTES

1. *U.P. Forest Statistics, 1981-82,* Published by U.P. government 1983, pp.194,198.
2. *Ibid.*
3. A.E. Osmaston *Working plan,* p.97.
4. H.G. Walton, *Gazetter of Garhwal Himalaya;* p.19.
5. Neeru Nanda, *Afforestation Strategies in the Bhagirathi Catchment,* Study conducted for National Centre for Human Settlements and Environment, Bhopal, p.58.

6. N.C. Saxena, *Farm Forestry in North West India: Lessons from the 1980's* Ford Founation, New Delhi, March 1990.
7. Anil Agarwal and Sunita Narain, *Towards Green Villages,* p.13, Wrong Trees or Dead Trees.
8. Indian Institute of Public Opinion, *Tree Survival Survey of Five States,* 1987 (unpublished).

CHAPTER 9

The Strategies of Plantation and Protection

FAILED PLANTATIONS

I visited Tamyar and Odada in March 1987. Odada village is situated off the beaten track across the Hyunl river and can be reached by a 24 km trek from Adwani which is located off the Tehri highway on the Gaja road. The sites covered by these tours were in the deep interior, away from the pull of urban centres and hence, strictly speaking, did not fall in the roadside "no-man's lands", since community sanctions were effective and the plantations remained undisturbed by the villagers. The total failure of these plantations therefore has to be attributed to other factors.

The first plantation of Odada was taken up in 1983-84 after which three more sites were handed over to the plantation division in two successive years. The plantations which I visited were widely scattered along stony, dry and barren slopes with giant euphorbia towering contemptously over stray seedlings barely an inch or two in height. Most of the pits lay empty but natural shrubs of tungla and karonda had benifitted by the enclosures and were well-looking, large and bushy. In one plantation there were two mother trees of chir pine which had shed a generous amount of seed in the enclosure, as a result of which rapid and prolific natural regeneration was taking place. While the planted seedlings bore a dismal look, mother nature was apparently able to do a much better job than the sarkar, without the sound and fury that characterises government efforts. The only plantation where seedlings had survived well had a medley of chir (planted in polybags), bamboo and acacia molissima. Its success was largely due to a nearby gadhera (mountain stream) where the pradhan had set up a farmers nursery to service the plantation. The department had helped him by providing free polybags and seed under its scheme of farmers nurseries, and so the plants were saved the harrowing trip up from the forest nursery, located near the river bed at Hadisera, from where seedlings used to be brought stuffed in sacks, with naked roots sticking out and jumbled up, swaying on mule back up a three and a half hour hot, steep climb to Odada.

It is not surprising that the plantations wore such a dismal look, considering the distance of the forest nursery, the poor size and growth of seedlings planted and the mode of carraige. Further, there was not even a semblance of an area plan for the village commons. The department started its work each year afresh, ignoring the previous years plantations as also the previous years failures. The officers (busy and harrassed with the urgency of meeting up with new targets) were mostly unable to visit old plantations which were simply forgotten once the target of plantation was achieved for that year.

The fact that old plantations are no where in the scheme of things was aptly brought out by a major flaw in the system of watch and ward which was pointed out by the Odada pradhan. Odada village was graced with the presence of four poorly paid chowkidars for four different plantations. The chowkidars for the 20 ha sites were paid Rs 175 p.m., while the others got Rs 100 only. However, cumulatively, the department was spending Rs 625 per month for chowkidari for four sites, an investment which was benefitting neither the villagers nor the plantations nor the chowkidars. One chowkidar with a living wage of Rs 600 p.m. might have managed watch and ward effectively for all four sites. A further problem (which was common to the whole district, if not to the entire state) was the fact that pay packets of chowkidars appointed by the forest department in U.P. hills used to arrive only once a year. (They could count themselves lucky if the money arrived every six months). The chowkidars salary had to be drawn from Bemunda by the forest guard at Jajal and then transported to

Odada. Which chowkidar would be motivated to work at such an irregular pittance ? But when it came as a lump sum it was naturally pocketed happily.

Odada pradhan suggested that chowkidars should be paid by the panchayats and also be vested with powers to fine encroachers. A serious weakness of the programme lay in the fact that the forest department had no legal jursdiction in revenue lands in U.P., hence the department, which invested crores of rupees in these lands, had no legal authority to protect the assets thus created. The plantation staff were unable to take prompt action against encroachments which had to be reported to the land revenue officials for taking appropriate action under the land revenue act. For each offence the forest guard had to search out the local patwari, lodge a report with him and then hope that he would follow up with the revenue courts which operate from the distant district headquarters and are already overburdened with multifarious litigations. Experience has shown that only the most dogged and determined forest guards were able to use the provisions of the land revenue act for punishing transgressors and if at all this happened, it was invariably after the event. Generally, forest guards avoided recourse to the land revenue act since the process was time-consuming and results were quite uncertain. Preventive action was not possible and if nothing emerged from his reports, besides losing a few seedlings, he would also have lost face before the people. Since grass-root extention workers survive on their status, however nebulous, in the eyes of the common man, and losing face is, in any case, a serious matter in oriental societies, the forest guards generally preferred not to report cases to the patwari. By and large, instances of encroachment were tackled through the village community and wherever the community leaders were active and able to control users, safety of plantations was ensured.

The Face of Success

At the cutting edge, the forest guard is the key functionary, who actually implements plantation schemes and interacts closely with villagers. For them the forest guard or patrol (as he is generally called) personifies the department. With him it sinks or swims. A sincere forest guard would have to struggle constantly to protect plantations set up by him. Though gaon pradhans used to be forthcoming with prastavs (proposals) for gherbar (enclosures), they tended to lose interest once the plantation work was wrapped up. Villagers whose kheti (cultivation) adjoined the enclosures generally made attempts to continue using the area for grazing or for cutting grass and the community could not always control them. The struggles of Jaikishan, forest guard Ghuttu provide an interesting glimpse into the ground reality of protection after plantation.

Jaikishan, who had lovingly reared many successful plantations in the Ghuttu vallley, was one of the most remarkable forest guards I met with during my travels. There were two villages in Jaikishan's area which had proved a source of constant harrassment. He had put in alot of effort to persuade the owner village to close the area which had subsequently shown radical improvement in biomass and density. All efforts went waste however, due to intervillage disputes which he was not empowered to solve. Ultimately, the rival party descended on the plantation and simply razed it to the ground, while Jaikishan remained a helpless, if pained, witness to this senseless act.

In Chakkargaon, next to Chetvan Gaon, Jaikishan had set up another good plantation which was encroached upon by a farmer who extended cultivation right up to the boundary wall and became bold enough to let his goats loose to graze inside the enclosure. Jaikishan repeatedly caught him in the act but met with only insolent defiance. He approached the pradhan for help, but came up against a wall of indifference. His protests went in vain while the goats continued to graze in the plantation, right under his nose. This was an insult which Jaikishan found hard to bear and which he could not instantly avenge. The whole legal process of the land revenue code was too lengthy and cumbersome, and the goats would have completely destroyed his plantation by then. Jaikishan was a man in a hurry, being too intense for methodical, snail paced proceedures devised by bureaucrats. Embittered by previous unhappy experiences and determined not to be a helpless witness any further, he decided, in a characteristically stubborn outburst, to adopt the time – honoured Indian technique of satyagraha. So, early one morning, he proceeded to the village square and sat down there in dharna, going on hunger strike before the whole village. As dusk gathered and villagers returned home, they found the young Gandhi sitting cross-legged, long-faced and obdurate, though hungry, in the public square.

Gandhian tactics certainly worked. The village elders gathered immediately and held a meeting in which they summoned the offender who was sternly upbraided and fined. They then came and apologised to Jaikishan and begged his forgiveness. "But I did not let them off so easily" he said, his face glowing pink as he added with gentle satisfaction "I took a guarantee from them written that this sort of encroachment would not take place again". The word got round and goats never entered plantations in his beat again, even though he did not exactly earn kudos from his superior officer for having taken recourse to such unorthodox methods to protect his plantations.

In the Ghuttu valley, a one day trek took us through some beautiful countryside and four excellent plantations set up by Jaikishan in Bhergaon and Medhu gram sabhas. The Bhergaon plantation was doing very well because villagers were sensitised to the need for eco-protection. Though there were an average of four cattle head per household, they had agreed to keep their cattle away from the site and had also sub-divided a portion of the common land amongst themselves for rearing sailma (thatch) grass.

The last and highest plantation site was at Medhu Kanyaj — a four km trek up the left bank of the Bhilangana through rather splendid scenery marked by huge bushes of rosa muscata in full bloom and enormous pangar trees with clusters of large white flowers. Enroute there were some good plantations also with chir, sheesham, pangar and walnut thriving and seed being dribbled into the pits where the seedlings had dried up.

If all government servants are really as corrupt as they are made out to be, why should the staff have bothered to lovingly rear saplings in such a remote site as Medhu Kanyaj? VIPs would never visit here in the normal course.

Thus successful plantations could be found in relatively isolated villages wherever the village community and a good forest guard or ranger joined hands.

Haridutt Semwal, Forest Guard of Dhung Mandhar (also in the Bhilangana valley), was less dramatic than Jaikishan, but an equally dogged and dedicated forest guard who had taken up the cudgels against encroachers through more conventional, bureaucratic methods. Semwal turned out to be quite the opposite of Jaikishan, a tall, quiet mature man of few words. He had served in this remote and arid area for three years and his experiences echoed the same signs of strain and lonely struggle which were undergone by Jaikishan.

Garh villagers had cooperated very well with Semwal. It had not been easy to persuade them to close their grazing area to cattle and wait for the grass to grow for a collective annual harvest. But they had listened to him and interest in closures had reached a peak when they saw the grass grow thick, tall and inviting amidst the planted saplings.

There is a saying in Hindi that a neighbours jealousy should always be feared more than a poisonous snake. In the Garhwal Himalayas this certainly proved true in the drought year of 1987. Seeing the lush growth of grass on the hitherto barren and neglected plot, villagers of Dhung suddenly revived ancient usage rights. Perhaps, because of the drought, they just could not contain themselves when the barren land enclosed by Garh suddenly became green and bountiful, while no grass was to be found anywhere else. One day the Dhung people entered the plot and forcibly harvested the grass, in the process damaging a large number of saplings. Semwal, however, did not let them off the hook and relentlessly pursued the case with the revenue administration.

The Mandhar nursery (which lies about 10 km uphill walking distance from Ghunti bazar) came as the culmination of a medley of remote Himalayan surprises which I had encountered so far in my travels in search of a forest. In the midst of an isolated divine, dark deodar forest, Semwal had successfully grown over 10,000 oak saplings all looking sturdy and proud in their poly bags. Since oak is a shade lover, this patch of forest was ideal for speedy growth of young seedlings. Infact, in the Rani Chauri campus of the G.B. Pant University, a young scientist I knew had proved through experiments that oak seedlings developed faster and better when put under shade, be it even with simple white markin cloth. I told Semwal about the same scientist's successful experiments for promoting quick germination of acorns by creating a greenhouse effect under polythene sheets during winter and he promised to try it out.

While the Bhilangana valley had received some rainful during August, Semwal's area did not get a drop of rain after June. Yet I was astonished to find that all his plantations were a tremendous success. When questioned closely he explained that he achieved success by raising saplings in a well-irrigated nursery located near the plantation site, and by invariably transporting and planting the seedlings in poly bags and not by the naked roots. Hence, in that dreadful drought year, Semwal had managed to ensure 75% survival of saplings — a truly remarkable achievement. Somehow I can never forget him as a stood, tall grave and quiet amidst giant euphorbia, saying (more to himself than to me) "I have tried so hard to make the villagers realise that our department is doing all this only for their benefit. These saplings now constitute their wealth. If only they can take care of it and nurture it, they will prosper."

Dharkot range which covered 45 villages located in a remote part of Pratapnagar block, had the maximum number of successful plantations to its credit. The entire range, which fell largely in an area of poor and degraded soil conditions, was known for this achievement. All over Pratapnagar block one could hear people talk solmenly, a little enviously, about effective plantation strategies in Dharkot range. Perhaps because the Range officer resided there and exercised careful control and supervision, flourishing plantations could be seen over an area of 1200 hectares which had been enclosed over a relatively short period of three years. Some landslip zones had been treated so ingenously (with a combination of bunding, soil conservation measures and plantation) that the problem sites could no longer be detected. No doubt, bhimal and dainkan had dried up even here, but chir in polybags had been replanted and ash, salex and robinia psuedo acacia were showing satisfactory growth. Hence, though the women toiled still, coming down, bent double under huge bundles of green grass brought from the peaks above Dharkot, these plantations provided atleast a ray of hope for their daughters and grand daughters.

All these were remote and interior sites. Along the roadside and national highway, there were only a few such sites, the most impressive being those of a rather unique roadside village, Patagli, located just above the main Tehri-Ghunti road, about 15 km from Tehri. In Patagli, common lands were strictly regulated and villagers cooperated to protect a total of four departmental plantations which were all thriving, not only in terms of rootstock but also by way of planted saplings. Another stray case was that of the Duadhar oak patch, also in civil soyam, along the Narendranagar-Agrakhal main road, where villagers had closed degraded oak areas on their own and handed them over to the department for plantation. But invariably, in these cases, the civil soyam land was not subject to external influences and was seen to fall very much within dejure and de facto ownership of the user villagers. The spindly, overlopped oak trees had benefitted greatly by enclosure in Duadhar and this example was being emulated in nearby Saldogi where villagers had handed over an oak wood to the civil soyam division for protection and plantation.

What did these successful plantations have in common ? With a few exceptions, by and large, the model outlined by Chambers, Saxena and Shah (quoted in the previous chapter) does explain the difference. Generally, it was seen that community responses in civil soyam land were positive in relatively remote villages where over 50 per cent of the families were dependent on agriculture and animal husbandry, and the villages had been able to retain the exclusive character of their civil soyam lands, either by virtue of being single owners (in case of a large village) or by subdividing clearly demarcated areas between three to four villages where more than one village settlement existed in a single gram sabha. In this way each village had been able to manage its own common lands, since they did not have to face external forces impinging upon village commons elsewhere and they were assured of equitable distributions of the usfruct amongst themselves.

In these areas the forest department obtained maximum participation of the people and overall success given the essential element of sincere field staff, good supervision and carefull follow-up. Generally, the surviving plants showed a preponderance of salex, robinia pseudoacacia, bamboo, chir, sheesham, cupressa, deodar, walnut and horsechestnut. But while stray NGO success stories get wide media publicity, these success stories of the forest department went generally unreported, perhaps precisely because of their remote location-remoteness from roadside and urban centres being itself the major factor in their success.

In many cases, villagers take initiative to close, protect and manage forests falling in civil soyam lands by a common resolution of the gram sabha and by appointment of a village chowkidar (who was generally paid Rs 500 per month in 1987). In these cases, there was no cooperation from the forest department since no plantation was undertaken. Extensive documentation of these villages will be presented in Part III.

To sum up, evaluation of plantations showed that plantation technology generally adopted was too primitive to be able to successfully manage the acute xerophytic conditions prevailing in most civil soyam lands, hence the departmental plantations showed best results in areas where some forest cover already existed, even though in a badly lopped and degraded condition. These conditions were rarely available in the civil soyam. Sites totally devoid of tree cover and infested with weeds and cactus had high failure rates. These sites actually needed closure and rest for a couple of years in which they could develop as good grasslands, yielding a useful annual harvest of grass. Attempts to bring them directly under tree cover invariably failed. In these sites even protection did not help in raising the survival rates, though the natural rootstock did gain considerably in strength and villagers also profited from enhanced yield of grass.

No doubt the plantation strategies of the department were largely to blame. Not only were lacs of seedlings planted by the naked roots, they were also reared in nurseries for five to six months only and were generally not mature enough to withstand the stress of uprooting, transportation over long distance by truck and head load, and subsequent transplantation, all without the cover and protection of a polybag. Polybags could not be provided because the extra cost would have increased plantation and transport expenditure, requiring reduction in the overall plantation targets fixed for the district, as also the target of 1400 plants per hectare for each plantation. Distance of the plantation site from the nursery was an important variable, the closer the nursery, better the survival. Thus hectic plantation targets swept aside all other considerations, eventually nullifying the very purpose of the programme. Conifers which have high survival rates were not planted in significant numbers due to local resistance. Salex, alders, sheesham, walnuts, horsechesnut, deodars and cupressas were all hardly suvivors and relatively fast growing trees. But even these species were not selected for planting in large numbers. Oak, a slow growing tree, much valued by the locals, was hardly raised at all.

Even if successful plantations are raised, the problem of security continues to plague them. Though they are generally respected by the villagers, they are liable to be destroyed by a single stroke of the administrators pen if parcelled out (as happens so often) to various public departments (or even to NGOs) for various public purposes such as setting up of schools, farms, health centres, hotels, not to mention allotment of leases to favoured persons for residence or cash crop farming or horticulture. No doubt, land with standing trees is protected by law, but its status as sarkari land in Tehri and the small and scattered nature of successful plantations ensures that they generally fail to show up as wooded land in the records and are lumped together with other degraded lands of the commons. Infact, when the tragedy and travails of civil soyam lands are seen in their entirety, it becomes difficult to comprehend how such a programme, crippled as it were from inception, could have dominated the scene for over a decade with funding to the tune of several crores of rupees, not only from the Government of India, but also from the World Bank and the European Economic Community.

No serious ongoing evaluation of strategies was undertaken by the state government and responsibility for poor performance was generally fixed on poor implementation by forest officials, rather than on short sighted policy planning by senior bureaucrats. The plantations policies of hop, skip and jump, planting crores of saplings any how and everywhere, had no admirers amongst either the villagers or the field staff of the forest department. If the hill forests were being restored anywhere it was happening as a result of quiet struggles and sacrifice of villagers or forest department staff or both together. These successes have taken place inspite of the constraints of target oriented policies dictated from above. There is no shortage of goodwill and money, but knowledge of community interactions at the ground level is lacking. A sustained perspective of protection and management of community resource banks over a period of 25-50 years is required. Government must accept the need for extended area protection and treatment, particularly of declining broad leaved forests, as against hectic target oriented plantations. The latter has almost reached saturation level. As one forest officer put it to the writer "The stage is coming when I will be planting gulmohars on Khatling glacier".

Could a better strategy be evolved for greening the Himalayas ? In the following chapter, various policy options are examined.

CHAPTER 10
Policy Options

While the failure of social forestry in village commons is more or less well known, if not intensively documented, it is yet too early to assess whether state policies have officially accepted the grim reality of its poor performance. In the sphere of family planning, the fiftieth year of independence has been marked by increasing emphasis on holistic mother and child care, with a concomitant move away from the earlier pre-occupation with fulfilment of sterilisation targets. The Ministry of Health is perhaps well on the road leading away from quantitative targets, but it is too early to say whether the Ministry of Environment and Forests is in a position to make a similiar break with the past.

The problem with evaluation and reformulation of afforestation policy also lies partly in the fact that there is an ongoing debate regarding appropriate strategies, some of which are pulling in opposite directions. There are no commonly perceived solutions to a universally perceived problem, perhaps because there are divergent views, not only with regard to the causes of deforestation in the Himalayas, but also concerning reasons for non-performance of the programme. Hence several, often opposing, ideologies and approaches for afforestation are championed and continue to operate simultaneously at different levels.

Farm Forestry and Tree Pattas

According to one viewpoint, the failure of the civil soyam plantations should be attributed to the collapse of the village community. It is therefore presumed that wastelands can be greened only by harnessing the enterprise of individual farmers for farm forestry on private land. Besides encouraging investment in farm forestry on commercial lines, this policy option favours parcelling out village commons to farmers for growing trees, either for subsistence or for profit. Supporters of this policy believe in actively utilising the profit motive of individuals or industry for restoring the Himalayas.[1] The force of logic could extend to a concomitant stress on eliminating community ownership in the village commons or even to lease of reserve forests to individual farmers, multi-nationals or business houses.[2] So far, however, it has been confined to state government tree patta schemes which give leases to farmers for growing trees on gram sabha land. Ironically, such well-meaning schemes for reafforestation have a conceptual and historical continuity with state policies of expropriation, insofar as they also prefer to deal directly with a multitude of individuals, facing either (a now enlightened), state or its fashionable alternative, the urban-based NGO.

Chambers, Saxena and Shah have argued strongly in favour of the tree patta scheme. While acknowledging that the scheme has not been a success, they lay the blame for failure at the door of bureaucratic and legal controls and present strong justification for abolition of Tree Protection Acts and liberalision of controls on felling and marketing of commercially valuable trees growing on private/tree patta lands. When applied to Uttarakhand the supposition would be that full market freedom would create the right climate for widespread greening of the Himalayas by allowing maximum profits from sale of valuable timber trees like deodar chir, blue pine, tun etc.

The case against privatisation of biomass resources has been ably presented by Anil Agarwal and Sunita Narain in their thoughtful monograph on the greening of India. While reviewing the performance of this scheme in Rajasthan and West Bengal, they have concluded that tree pattas tended to dramatically increase conflict situations within the village, even when distributed with care to deserving persons. In Rajasthan they found instances where poor families had been socially boycotted after they meticulously closed and protected of common grazing lands alloted to them under this scheme.

In West Bengal the response to farm forestry was particularly encouraging, and, at one point, the government seriously thought of also leasing out degraded forest land to individuals for purposes of farm forestry. However, since the degraded forest area is around 0.65 million hectares, even if two hectares of this land was alloted to each family to grow trees, *only 4 percent of the total rural families would benefit, while the remaining people, all dependent on these lands, would lose out. This situation would obviously create immense conflicts.*[3]

It cannot be disputed that equity will be the first casualty of widespread moves to privatise the commons by granting tree pattas which are also likely to be monopolised by influential members of the village community. In Tehri these pattas have been used largely as a means for legalising encroachments on civil soyam lands, and can only result in further reduction of the biomass available to deprived sections in the village. In Jakholi block, many villages have, by and large, subdivided the common lands which are being very well protected by individual farmers since many years. However, even in these villages, a handful of families who are left out have, of late, become rather vociferous in expressing their resentment. In case the state permits commercial marketing of the trees which are at present only being used for fuel and fodder, these suppressed grievances are likely to flare up.[4]

In a discerning article on the subject, Shashi Kant has analysed the whole policy for privatising the commons for tree farming. He has questioned the assumption that natural productivity of common lands will increase by parcelling it out to individuals. He has shown how administrative costs of allotment, transfer, demarcation, fencing and protection of individual plots along vast stretches of village commons will actually be counter productive. Such transfers will tend to establish a bureaucratic stranglehold over the village scene, since all this parcelling out of land and the accompanying disputes will have to be handled by the revenue administration, for whom such a policy (if implemented on a large scale), would be a windfall indeed.[5]

Shashi Kant has further shown that the per capita availability of land would be 0.2 ha only if this exercise was carried out equitably. 19,276 ha of land was covered by tree pattas in U.P. by March 1988. A minimum investment of Rs. 15000 per ha would be required to treat these highly degraded soils with sophisticated technology. This would require a net area of 5 ha for the land holding to be cost effective. Against this requirement the average size of plot per patta holder works out to 0.32 ha. This is not a viable proposition.[6] Such holdings would be unviable, serve no purpose and give rise to innumerable conflicts between individuals who are residing as members of a community which may not have strong control over its village commons, but still does identify as a socio-geographical and cultural unit. The policy of privatising commons, while recognising the malaise, ends up by throwing the baby out with the bath water.

Perhaps the opposition between commercial forestry by individuals as against forestry for sustenance by village commons is more apparent than real. Since the commons in U.P. Himalayas and particularly Tehri Garhwal are generally in an extremely degraded state with poor soils, arid conditions and scrub vegetation, it is a question of whether higher productivity will be achieved on these degraded lands by management through individuals or by the village community. It goes without question that the size of unit of production will decrease with the privatisation process and therefore economies of scale, which govern profitability, would be adversely affected.

Fifty years of independence have consistently shown that wherever land with any kind of economic value has been distributed to the poor, (whether in slum resettlement colonies in urban metropolis or to tribals in remote, densely forested Bastar) land grabbers have been effective in buying out the poor and acquiring economic control over large areas which have got progressively alienated, despite social legislation to the contrary. In an important and widely quoted study of dry villages in India, N.S. Jodha found that when community lands were privatised, over half the area in his sample study found its way into the hands of the non-poor sections.[7] This is a process which has been going on since the colonial period and free India has been unable to reverse it. There is no reason to presume that events would take a different course in the Himalayas. Infact, a great tragedy would have overtaken the region had restrictions on tree felling and marketing been withdrawn. The tree patta schemes would have become instrumental in virtually parcelling out civil soyam lands, eventually also threatening the reserves. Since a sizeable portion of civil soyam is held under thirty year leases by farmers, and some tracts are wooded, removal of restrictions on felling and free movement of marketable trees might have led to large scale deforestation, not only of private woods, but also of woods on civil soyam lands held on patta, and other commons just adjoining the reserves where the boundaries can be considered to be indeterminate. The fact that the law pertaining to tree felling was not changed, is certainly one of the reasons why the tree patta scheme did not become commercially popular. Had these laws been relaxed, urban based property dealers and land grabbers could have made all-out efforts to appropriate the civil soyam lands, and also parts of the reserve forests, in a big way. These apprehensions are now being confirmed by increasing appropriation of land along the Dhanolti Chamba road by hoteliers and tourist operators. Similiar alienation of land seems to be taking place enroute to Har ki Dun along the Mori-Naitwar, Taluka-Osla axis.

It would be patently unfair to suggest that the advocates of the tree-patta and farm forestry schemes have anything but the genuine interests of poor farmers at heart. But the policy is not pragmatic for reasons explained above. At the most, there is some scope for farm forestry on private fields. The household surveys of Dang and Bel undertaken by the author in 1987 also indicated that villagers are already growing sizeable number of fuel and fodder trees in their fields at an average of about 30 trees per household. Hence farm forestry with native fuel and fodder trees is an acceptable proposition within the overall constraints of availability of land.[8]

If we accept the option of retaining the civil soyam as a

common property resource, what are the policy alternatives available for the Himalayas? The earliest option is that of Van Panchayats which were set up in British Garhwal (but not in Tehri) subsequent to the Kumaon Forest Grievances Committee 1922.

Van Panchayats

The area covered by these panchayats is given below.

TABLE 1

Area under Van Panchayats in Uttar Pradesh Hills (March 1982)

District	*Total Forest Area (ha)*	*Van Panchayat Area (ha)*
Almora	394,400	62,800
Pithoragarh	330,300	71,400
Nainital	420,600	20,800
Garhwal	455,600	29,600
Chamoli	526,900	52,100
Total	2,127,800	236,700

Forest Statistics of U.P. 81-82.

Van panchayats have had a chequered career. Many are so well-protected that all extraction by villages for fuel and timber is directed towards reserve forests which continue to deteriorate, though it must be admitted that forest area per se is preserved, even though it is at the cost of the state forests. Strangely enough, both van panchayats and, of course, the reserves, have remained neglected by way of investments for afforestation, while heavy investments were directed towards civil soyam lands.

Only a few panchayats, some of them in Landsowne area, are known to follow management systems whereby blocks or compartments are worked in rotation within the overall area available with them. In the border subdivision of Dharchulla, van panchayats appear to be functioning better than in more developed areas of Nainital, though it must be said to the credit of villagers that certain tracts, as in the Gairsein block, show remarkable crown density and are in prime and pristine condition with which the average reserves cannot compete in any way. Chamoli district is also distinguished by excellent van panchayats but, by and large it is the older ones which are healthy and stable. During the seventies there was a commendable effort to form more van panchayats in Kumaon, but by then the pressure on civil soyam lands had come to stay, and the formation was either retarded by inter-village conflicts or the panchayats were constituted mechanically, just to fulfil targets. Many of them were of size so small as to render them quite insignificant.

Some van panchayats have followed the profit motive as a result of which, in stray cases, trees have been commercially felled or resin tapped, sometimes excessively, at the instance of unscrupulous pradhans. But the Van Panchayat Act provides for only a proportion of the revenue to be given to the village account, the balance accruing to the forest department. With current policy moves to give the entire profit to the van panchayats, the potential conflict between biomass needs and revenue demands might increase.

Management of van panchayats is also progressively being seen in commercial terms. Success is measured, not by the degree of protection or crown density achieved, but by the amount of revenue which has accrued to the panchayat fund. The move to give the entire revenue earned to the panchayat is based on the supposition that once communities are granted these rights and given an economic stake in the forests, they will surely unite for better management and protection. The fact that these notions are somewhat euphoric emerges from sincere field studies such as those undertaken by Agarwal and Narain, who found poor villagers in Maharashtra quite alienated from the panchayat lands. None of the panchayat leaders had given much thought to equitable distribution of the biomass and most of them were excluding the poor and auctioning off the grass to contractors.[9]

The formation of van panchayats in Tehri district is a welcome management option, but by the very nature of the scheme they could only be set up, by common consent, in the least problematic lands which are free from disputes. They will not provide any startling solution to the problem, since they are not likely to succeed in the no-man's lands, and will face the same conflict situations and similiar biotic pressures and problems posed by multiple users. The lands are so degraded that the profit motive will not work here in any case - protection for biomass production in terms of grass and shrubs is all that can be hoped for.

The second defect is with regard to management. Formation and management could have vested with the forest department, but, as matters stand, formation of van panchayats is also not an easy task. Odada pradhan had been trying hard to set up a van panchayat for a sal forest patch which the villagers were protecting in common land, but he faced endless problems since the file on formation kept being tossed around between three offices — revenue department, panchayat directorate and forest department. As of today, in 1997, all out efforts are being made by the revenue department to set up van panchayats in civil soyam land in Tehri district and about 161 panchayats have already been constituted. However, the exercise seems to have been undertaken more or less on paper, in tandem between the patwari and the pradhans, while the common people seem to have remained generally unaware of the whole process.

As will be shown presently, the Panchayati forests in Nepal suffered a similiar fate, but the government there was fairly quick in realising the problem and formalising alternatives. The previous chapter on plantation and protection strategies has also revealed the serious handicaps under which the forest department staff have to function in civil soyam lands which vest with the revenue department. Unfortunately, van panchayats in Tehri are being set up in these very civil soyam lands which are generally wracked by disputes and have, as key managers, the revenue staff, for whom setting up of van panchayats is just another task to be performed amidst their multifarious duties. It is not surprising that the panchayats have been formed mechanically, in most cases without even the general body meeting stipulated in the Act.[10]

The interaction of the van panchayat management with the

existing political framework of panchayati raj, party based politics and other village institutions has proved to be occassionally problematic. The writer's personal experience of introduction of mainstream democracy into the remote mountain area of Tawang, in Arunachal Pradesh, has shown that, while the hill people take to our form of politics like a duck to water, our democratic structures have completely undermined their traditional systems which had once existed for management of co-operation and conflict. The functioning of our democratic institutions seems to have vertically split the villages in Tawang where people now occassionally indulge in factional politics even for such emotive matters as repairs of the village gompa (temple). The consensus and unity for management of community property resources appears to be fast receeding in these mountian communities.[11]

After extensive travels throughout India, Anil Agarwal and Sunita Narain concluded that, in almost all cases where voluntary agencies have had to organise communities to manage their natural resources, they have had to bypass the formal village institution of the panchayat. Their findings, which are based on a cross-section of experience, exactly tally with the writer's observation of Tawang over a period of two decades, namely that village panchayats are both the products and the promoters, of village factionalism—the very antithesis of the unity which was the hallmark of indigenous management systems which Mahatma Gandhi ideally conceived as village republics.[12] Besides this, panchayat areas of jurisdiction are not coterminus with homogeneous village settlements. The problems which arise as a consequence have already been described at length in a previous chapter.

Van panchayats cannot be considered the last bastion of community resource management in the Uttar Pradesh Himalaya nor can it be said that they have even remotely reproduced the indigenous management structures which had existed in the Himalayas before the onset of the Raj. The picture on the ground is more complex. For this reason, details of van panchayats established in Tehri Garhwal in the decade of the nineties are given in the last chapter of this book, after presenting extensive documentation of the ground situation with regard to indigenous management systems existing as vibrant, effective and parrallel structures to the formal government institutions.

Management of community forests in Nepal

We have much to learn from the Nepal experience, since HMG Nepal has made repeated and bold changes in state policy in its efforts to obtain peoples participation for sustainable management of the nations forest wealth.

While the forests of the terai used to be considered as the property of the rulers of Nepal, they had not interfered with the upland forests where the hill people had traditional established systems of managing their forest wealth. These involved locally accepted rules through which a clearly fixed group of beneficiaries regulated forest use and excluded outsiders. These local systems were recognised by the Rana Government. After the overthrowal of the Ranas in 1950, the state decided to nationalise the forests in 1957. Predictably, this led to widespread mismanagement and destruction, because, in the words of K.B. Shrestha, Deputy Director General Community Forests Nepal, "the local people lost all benefits from the forests and over time became apathetic towards government because they had been deprived of management rights and benefits...*alienation of people from resource management brought about wanton destruction.*"[13] In order to rectify this, HMG Nepal created panchayati forests in 1976 but this venture also failed. The forest area of the country decreased from 6.5 million hectares in 1965 to 5.5 million hectares in 1990. Entrusting the local resources of the community to politically elected bodies was found to be ineffective after about a decade. One of the important reasons (which has also been found to be true of the Indian situation), was that the elected panchayat was found to comprise too large a unit to develop genuine interest in supervising and managing the local forests.[14]

In 1990 fresh efforts were made by HMG Nepal, this time to hand over forests to identified user groups for management and usufruct, subject to an operational plan approved by the DFO. This policy was formalised by the 1993 New Forest Act of Nepal, an extremely progressive legislation not very widely known in U.P. hills. Under this Act, the DFO can hand over a tract of forest directly to user groups, who are empowered to price and sell the produce, subject to an approved operational plan. While the handing over of natural forests to user groups has proceeded more or less smoothly, even here the profit factor has tended to give rise to conflict situations, since trees are often cut and sold in violation of the approved plans, as a result of which punitive action has to be taken by the department. Protection and extraction for biomass by user groups has so far shown more promising results than plantation for sustainable management and this unique experiment should be watched and monitored keenly by policy planners working for afforestation in the central Himalaya.

Ofcourse there is a vital difference between the situation in Nepal and Uttarakhand. Nepal has 75 districts with 75 DFOs to cover a land mass of 14.7 million hectares with 5.5 million hectares of forest and an average of only 50 villages per district. Compare this with Tehri district and its 1943 villages and 0.26 million hectares of forest area under a single DFO (an additional DFO without any additional staff has been provided only recently in 1996). The personal touch and close contact and involvement with the people required under this pattern can only come about if the management structure is radically strengthened, perhaps by merging all the staff of territorial and plantation divisions into several, single divisions with combined responsibilities over smaller and more manageable areas of jurisdiction.

The Nepal legislation is in tandem with contemporary developments in India which have been popularised under the framework and concept of Joint Forestry Management.

Joint Forestry Management

West Bengal was the pioneer in this field and still leads insofar as the state government is committed to bring all state forest lands under joint management by 1998. This policy option emerged in India after the success of the innovative Arabari experiment started indigenously by the West Bengal forest department in 1973 and popularised in other parts of the country in the next two decades.[15]

The sal forests of southern Bengal yield considerable amount of minor forest produce which is gathered and sold to contractors by poor and marginal farmers and landless labour who are generally exploited by the farmer. As explained in the first part of this book, the traditional system in Garhwal, and perhaps also in other parts of the country under native rule, had been to allow the villagers unrestricted rights of gathering forest produce which was later purchased by local contractors who paid an exit tax to the native rulers. This system was supplanted by the British penchant for dealing with middlemen who consequently proceeded to exploit the same villagers to gather the self-same produce, now as bonded or otherwise indigent labour. The fact that this new scheme was started in the communist ruled state of West Bengal is not without significance. For the first time after independence, the state sought to set the clock back and redress the wrongs of the colonial past. Since the whole system of auctioning out coups cannot now be reversed, the West Bengal forest department hit upon the idea of "returning" a portion of the revenue to the villagers by giving them a share of the profits (upto 25% in disposal of poles and timber) while they wisely waived all fees for gathering minor forest produce. In a limited way, this was a return to forest revenue policies under native rule and it was done specifically to enlist co-operation of the people (mostly landless labour or small and marginal farmers) for protection of sal forest tracts which are frequented by these poverty-stricken villagers who gather the minor produce while also foraging for their daily headload of fuelwood.

Under JFM, certain rights are accorded to the users, concommitant with the responsibilities they have to shoulder for voluntary policing and protection. The fact of their rights being recognised, along with obligation to perform certain duties for the upkeep of the forest, is the critical factor in JFM, along with sharing of revenue between the department and the people. Perhaps, for the poor families in this area, the recognition of their right to minor forest produce and the waiver of fees was a far greater incentive than the prospect of revenue receipts from sale of timber. Infact, many Forest Protection Committees were seen to object to silvicultural practices which promoted pole crop but retarded proliferation of leaves and flowers which were seen to yield a higher and steadier income.[16] Another important intervention, which should not be lost sight of, is the responsibility which the Forest Department took upon itself to make up for revenue loss due to closure of the forest and ban on collection of its produce for the period of two or three years which the forests required for recoupment. It is significant that the people's own efforts, for a very similiar management system of protection, by village groups all over Tehri district, carry no such benefits, since there is no external agency involved in the process. Also, since the hill people depend upon their forests primarily for fuel and fodder, wage labour by itself cannot provide a substitute for the loss occassioned by closure and protection (as is the case in West Bengal).

Uttarakhand is unique in sofar as rights of villagers in the erstwhile native state have been guaranteed under the instrument of accession while similiar rights continue to be honoured in British Garhwal. Hence their dependence on the reserves for fuel, fodder and timber has state sanction. The problem here lies in the fact that village user groups would need to be officially invested with responsibilities for protection and management which they are already shouldering, on their own initiative, in the reserve forests. While they already enjoy rights of extraction for domestic use, they cannot allocate these amongst themselves. For instance, village communities which are voluntarily protecting large patches of reserve forests have no right to allocate timber trees to households under the rammanna system, which still subsists directly between pradhans and the department, whereas, normally, the community itself could be a fairer judge of individual needs and greater transparency would also be possible if the community itself managed distribution of trees between households. These rights are ofcourse, available to the van panchayats, but panchayat area is limited and even the panchayat legislation needs to devolve more power, as also revenue, away from the department in favour of the people.

In a series of discussions recently held separately with senior forest officials and women workers of the Mahila Samakhya Tehri (who are drawn from the grassroots) a remarkable divergence of opinion emerged with regard to collaboration in tracts being voluntarily protected by villagers in the reserve forests. While Anand Singh Negi, a dynamic and progressive Conservator of Forests, was not adverse to carving out van panchayats in the reserves, the sahyoginis or women workers (who supervise and guide about ten village sanghas each) were not in favour of changing the legal status of the reserve forests and were therefore more in favour of JFM.[17] While the department does not extend any special rights or special help to the women who organise themselves to manage forests located in the reserves, they also do not interfere or hamper their work. In the changing political scenario today it is ironic, but understandable, that the sarkar is more impartial, and the rule of law and the state more dependable than local politics and local politicians. No doubt we have to find the right mix in the effort to decentralise management and keep it as closely as possible in the hands of the users. But those of us who seriously contemplate complete abdication of state responsibility are perhaps beating a dead horse.

Voluntary Forest Protection

At the time this study was being conducted and documentation of peoples protected forests being undertaken by the author in Tehri in 1987,[18] similiar studies were also being conducted in other parts of the country and these yielded similiar valuable information with regard to voluntary forest protection by villagers in other states. Mark Poffenburgher and Mcgean have recently documented almost all the major studies available on JFM and voluntary village protection and the accounts given below are largely drawn from their recent and important publication on the subject.[19]

Studies indicate that tribal districts of southern Bihar have shown an increasing engagement with voluntary forest protection since the early seventies, though it is not clear as to whether the fact of protection came to notice in the seventies or whether the protection process started then. Certainly the Jharkhand movement has given an added impetus to these efforts. There are also other similiarities with Uttarakhand in so far as tribals have forcibly prevented commercial extraction in some areas, and that there is little or no interaction with the forest department at the time of organising for protection.[20] A survey of voluntary forest protection in Chotanagpur, conducted by Shivnath Mehrotra and Chandra Kishore, found thirty two communities involved in forest protection, representing about 15 percent of all villages in the area.[21] This has interesting parrallels with the writer's study of forty three communities in Tehri which represented about 10 to 25 percent of villages in the area.[22] As in West Bengal and Uttarakhand, these groups base their strategies on protection and regeneration, the emphasis in West Bengal and Bihar being on restoration of degraded sal forests through strategies of protection quite similiar to those being followed for preservation of oak forests in Uttarakhand.

The example of Orissa is even more interesting than Bihar, and seems to come closer to what has been happening in Uttarakhand since traditional systems of governance appear to have survived amongst tribals and non-tribals in far greater strength in Orissa than in other parts of India. One of the reasons for this survival is interesting and signficant. Since the political system of panchayats in the state encompassed very large areas, comprising upto ten and twenty settlements, indigenous management systems retained their function, actively fulfilling the needs of sustainable management of community property resources. Shashi Kant has documented these systems and rightly pointed out their proximity with Gandhian approaches, thus incidentally confirming the strong pragmatic basis of so-called Gandhian idealism.[23]

The intervention of the state for spread of JFM in Orissa has some interesting lessons for Uttarakhand. On 1st August

1988 the policy of forest protection through community management was announced with fanfare and the department swung into action with the target of forming 5000 FPCs by the end of the year. An active and alert Gram Pradhan of Tehri who was much committed to tree farming had once remarked to the author "Yeh target bazi sub se badi bemari hai." (This obsession with targets is the greatest disease). Predictably, the forest department of Orissa achieved its target, but a majority of the newly formed groups remained non-functional while active, traditional groups were often left unrecognised. This dependence on state initiative is slightly different from the Nepalese approach where the user groups themselves request the department for recognition.[24] Still, by the end of 1993, about 27 per cent of Orissas forests were under some type of community control.

While Poffenburgher and Mcgean have described community efforts at sustaining forest wealth throughout the length and breadth of the country, the example of traditional management in the Western Ghats comes closest, like Orissa, to the Uttarakhand pattern. Madhava Gadgil has documented the indigenous management of sacred groves and community forests in the pre-British period, and recent surveys have also indicated that village resource protection systems are still operating in many areas. Out of 80 such groups managing their forests in Shimoga and Mysore districts, some were found to have been functioning since generations while others had started their work in the past 20 years. Data presented in part three will show how strikingly similiar these findings are to the documentation undertaken by the writer in Tehri Garhwal.[25] In other parts of the west coast, the communidades of Goa set up by Portugese rule no doubt undermined the quintessence of the native village management institutions of Goa, but the form and its funcitons were both retained, and throughout Portugese rule they continued to actively protect the common property resources of the village, the most important being the khazans which regulated the flow of fresh water into the village fields while preventing inundation by saline water, the whole process being managed intricately through the medium of extensive belts of mangrove reserves guarded voluntarily by the people.[26] Infact, the continental people of Europe seem to have had a more pragmatic approach than the British to the rural management systems which they found in India, and it is a pity that we ignored these small examples. Instead we are having to learn the hard way, so that fifty years after independence we are only now setting ourselves the task of reworking and remodelling the bureaucratic institutions which we received as part of our colonial heritage.

Forests for whom?

In the 1950s the policy emphasis was not just on industrialisation but also on growth and development of the state as an entity apart from the people. The vital importance of the Himalayan forest wealth for supporting local residents still needs to be better appreciated by policy makers, even though much water has flown under the bridge since then.

By providing for subsistence needs of the rural poor, the biomass yield assumes an economic or replacement value in terms of import of gas or firewood from the plains. One cannot escape the conclusion that the overwhelming emphasis on cash returns from timber or even minor forest produce stems from a purely male point of view. The women of Pratapnagar intermittently hang or drown themselves because of failure to reach home with their daily headload of grass.[27] Little five and six year old girls tramp long distances daily to forage for a headload of fuelwood to feed their fathers and brothers. Not a man is to be seen amidst the unending line of Pratapnagar women wending their weary way home, long after darkness has overshadowed the harsh and denuded mountainside. The responsibility of collecting fodder and fuel being vested with women, scarcity conditions, and the economic and replacement value of the biomass thus gathered need to be constantly kept in mind by policy planners. In a micro-level analysis of community property resources and the rural poor in 14 villages of Karnataka, Syed Ajmal Pasha found that around 10 and 6.2 percent of the gross income of poor and non-poor households respectively comes from community property resources. This calculation was made by imputing cash value to the fodder and fuel collected as also by giving an imputed value to the fodder grazed by the ruminant livestock. The writers have correctly noted that since the cash income of the rural poor is hardly enough to provide two meals a day, spending on fuel and fodder could be suicidal for them. Hence they have also supported policy options which protect and regenerate the CPRs while giving the poor increased access to these areas by providing equal opportunity to participate in their use and sustainable management.[28]

Even today, after all the awareness that is being created for gender sensitisation and after all the documentation and research on the economic value of fuel and fodder gathered from CPRs, sections in government are eager to push the case for Van Panchayats to grow geranium as an intensive cash crop to feed the perfume industry in the west. It is a cause for surprise and dismay that van panchayats are being urged to adopt this latest fad in the name of earning revenue.[29]

The question that arises is—revenue for what purpose and for whom? Will the money buy fuelwood and fodder? Policy planners should be aware of what has happened to the economy of African countries which had converted from traditional agriculture to farming for so-called high value cash crops destined for the affluent European markets. Is precious rootstock of high calorific value shrubs and grasses to be pulled out now and replaced by geranium so that the profit margins of western aromatic – based industry can be further maximised ? Who will replace the fuel and fodder being presently foraged from these lands? Who will pay for the subsidies (since the poor can hardly afford to pay) and how will it be transported from the plains? Such myopic policies can play havoc with the rural economy in the Himalaya in the name of making short-term quick profits.

For whom were the forests reserved in the first place? Was it not for the state, for industry, for contractors and businessmen to get rich quick? It was so in the days of Ms Reid and Finn and it seems to be the case today, whether for sale of timber from Arunachal or for import of timber from Malayasia, no matter whether it was to meet the needs of railway lines which had to be laid in the nineteenth century, or ship building during the war or for our paper and pulp industry after independence.[30]

Even when the ban on felling of green trees was imposed, for whom were the forests being conserved and protected? Surely for eco-protection, for the sake of the nation, for the good of humanity? If native people and their survival needs figured in this calculation, the whole scheme of biosphere reserves involving expulsion of indigenous people from their native habitat would not have achieved an international status with international funding. A detailed, section by section analysis made by Anita Cheria of the plan to set up Nagarhole National Park in Karnataka, with world bank assistance, makes interesting and poignant reading in this context.[31]

In the particular circumstances in which the Indian administration took over the mantle of the British Raj, radical thinking would not have been possible, atleast not in Uttar Pradesh. Only the fierce resistance of the chipko activists, when combined with the vision of Prime Minister Indira Gandhi, could result in the total ban on felling of green trees above an altitude of 1000 metres. This was a watershed which gradually led to an acceptance of the role of the people and their contribution towards preservation of forest wealth.

In order to have a mass programme for plantation, protection and regeneration for the people in the reserve forests of the Himalaya, the state would have to declare that the first charge on the forests of the country lies with the native people for fulfilment of their biomass needs. The protagonists of the people would also have to clear their heads of notions of making money out of forests and justifying forestry for profit simply by replacing the greed of the state by the greed or even monetary compulsions of local people.

In the ongoing debate of profit *vs* subsistence, N.C. Saxena has given perhaps the most lucid and balanced recommendations. After a detailed study of farm forestry in Uttar Pradesh he has concluded that the money economy and market forces govern farm forestry on private lands wheras the reserves continue to be drawn upon for fulfilment of biomass needs. Saxena, a very senior bureaucrat of the All India Services, recommends a reversal of the National Commission on Agriculture policies which regarded reserves as the source of commercial profit for the state and private and degraded commons as the biomass banks for growing fuel fodder species for the people. Instead he suggests that market demands may be met from private lands while changing the composition of monocultured tracts of reserves to a multi-layered multi-species forest area from where the poor can derive their subsistence needs, as also petty but assured income from minor forest produce.[32]

Well-meaning environmentalists may like to persuade the state to abdicate its responsibilities and hand over the reserves for distribution to the 'people', whether through Van Panchayats or Farm Forestry. However, no matter how impassioned one may be about the injustices of the past, ones prescriptions for the future should be guided, not by emotion, or even by ideology, but by the trend of ground reality which has been shaped and adapted by the people themselves in their quiet, unsung struggles for survival. That compulsions of eco-protection, ownership of the forest department and the biomass requirements of the poor can co-exist, and help in preserving the forest wealth, is an axiom that has to be recognised and actualised. This is what the forest department themselves discovered, experienced and popularised in West Bengal after the Arabari experiment.

Motivation to protect and restore the Himalayan wealth in extant common lands and reserve forests does exist amongst villagers, but profit by way of cash earnings cannot by itself provide the optimum conditions under which this motivation could be actualised and translated into effective protection.

We have to perhaps continuously remind ourselves that, historically, Himalayan forests of this country have constituted a biogenetic reserve producing 'minor' forest produce rather than timber, that they form an interconnected eco-system and living entity and cannot be parcelled out like agricultural fields whose records will be controlled and managed by some patwari or forest guard. Forests have to be loved and owned by communities who are held together by bonds of veneration and occasional fear that accompanies magico-religious perceptions of sacred spaces. Protection of village forests and indigenous management systems are tied up with the web of life in villages which have traditionally existed as a homogeneous unit, respecting sacred spaces, with common rituals and a shared perception of common threats and perceived solutions.[33] A dynamic community is an axiomatic precondition for survival or establishment of an indigenous management system which is based on subsistence for all and not on cash earnings for a few. Part III of this book will show how peoples efforts in inumerable villages quiet, steady, unsung and unhonoured—give strength and authenticity to these assumptions.

NOTES

1. Chamber, Saxena & Shah, see p.163 for 'Historical Overview of Government Land Tenurial Policy.' They rightly highlight the fact that Govt. programmes tended to be influenced by the idea that there are only two sustainable policies : either the commons should be privatised or they should be brought under the control of a coercive authority. Also see p.193, 'Various Hassles in Tree Patta schemes' for progress of Tree Patta Schemes in major states.
2. *Ibid*, see p.175, 'why small farmers don't plant trees'.
3. Agarwal & Narain, *Greening of India*, see pp.14-16, 'Privatisation or the Common way ?'
4. Personal Communication and Reports from Sahyoginis of Mahila Samakhya, Jakholi Block.
5. Shashikant, 'Privatisation of Common Property Resources, case of Tree Patta Scheme in U.P.' in the *Price of Forests,* edited Anil Agarwal, Centre for Science and Environment, New Delhi, 1992.
6. *Ibid.*
7. N.S. Jodha, *"Common Property Resources and Rural Poor in Dry Regions of India"*, Economic & Political Weekly 21, 1986.
8. Agarwal and Narain, *Towards Green Villages,* see pp. 21-22, 'Problems with Panchayats.'
9. Personal Communication to the Author during meeting with Mahila Samakhya workers in September-October '97. Mahila Samakhya has also published a booklet *"Jungle Aur Aurat"* in its series *"Bainyon Ka*

Raibar", in which the existing Van Panchayat Act & Rules have been given along with the draft Act & Rules according to proposed amendments. Despite this there appears to be considerable lack of awareness of the rules and the legal processes required for formation of Van Panchayas, not only amongst villagers and Sahyoginis but also amongst the revenue staff, who are responsible for the establishment of the Van Panchayats.

10. Personal Communication to the Author.
11. Agarwal & Narain, *Towards Green Villages.*
12. K.B. Shreshta, *Community Forestry in Nepal, An Overview of Conflicts,* Discussion paper series International Centre for Integrated Mountain Development, 1996.
13. *Ibid.*
14. K.C. Malhotra & Mark Poffenberger, (edited) *Forest Regenration through Community Protection: The West Bengal Experience,* West Bengal Forest Department, 1989.
15. N. Satish Chandra & Mark Poffenberger, *"Community Forest Management in West Bengal : FPC case studies".*
    U. Bannerjee, IFS, Addl. CCF West Bengal, *Participatory Forest Management in West Bengal.*
    S. Palit, Addl. CCF, *Present Status of FPCs.*
    All the articles listed appear in *Forest Regeneration through Community Protection., The West Bengal Experience,* 1989.
16. Neeru Nanda, "Local Management of Community Property Resources", Anil Agarwal (ed.) *The Price of Forests, CSE,* 1992.
17. Mark Poffenberger & Betsy McGean, (ed.) *Village Voices, Forest Choices, Joint Forestry Management in India,* Oxford University Press, 1996.
18. *Ibid.* pp.32-33.
19. Neeru Nanda 'Local Management of CPRs in the Himalaya in Agarwal (ed) *The Price of Forests* CSE 1992.
20. Poffenberger&McGean, *Village Voices Forest Choices* pp.33-34.
21. *Ibid,* pp. 34-35.
22. *Ibid,* pp. 43-44.
23. Shashi Kant, *op. cit.*
24. K.B. Shreshta, *op. cit.*
25. Neeru Nanda, *op. cit.*
26. Rui Gomes Pereira, *GOA: Gaunkari, the old village associations* edited and published by A Gomes Pereira Panaji 1981.
    Authors administrative experience as Secretary Forests Goa 1988 supplemented the accounts of management of mangrove forests for protection of khazans.
27. Mass suicides of young women have taken place in Tehri Garhwal due to extreme hardship. The case of three women of Khaspatti

who drowned themselves because they went to see the famous film *Ghar Jawain* instead of going to fetch their daily headload of grass has gone into legend and sunk deep into the psyche of women. Sunderlal Bahugana has often spoken and written about it. Mahila Samakhya *Rant Raibar* regularly highlights such cases.

28. Syed Ahmad Pasha, *CPRs and Rural Poor: a micro level analysis* Economic and Political Weekly 14 Nov. 1992.
29. Personal communication from the U.P. Academy of Administration who have brought out booklets to guide van panchayats on preparation of projects for plantation of geranium. It is obvious that, while one god fails, the establishment is not tardy in looking for other gods
30. The best analysis of state subsidised support for industry can be found in Madhava Gadgil, 'Perverse incentives for industrialisation and its impact on forest resources' in Anil Agarwal (ed.) *The Price of Forests* CSE 1992.
31. Anitha Cheria, *Why does Nagarhole Buru? Adivasi, the state and the World Bank* Edwin 26 VGP Shanthi Nagar, Pallikaranai, Chennai, India.
32. N.C. Saxena, *Farm Forestry in North West India: Lessons from the 1980s* Ford Foundation, New Delhi 1990.
33. The literature on sacred spaces is scattered but quite vast. The writer has explored the myths and legends which surround sacred sites across the Himalayas in the following articles and books:

    Neeru Nanda *Tawang, the land of Mon.* See chapter 5 for Gudpi, the mountain peak revered by graziers of Khet, chapter 10 for an account of the sacred spaces of Bangajanga, Takstang etc which were sanctified by visits and sojourn of Guru Padmasambhava and the account of Chongchongma and its transference as a sacred space from the Bon Gods to the new religion brought to Tawang by the great Guru. Chapter 10 also has an account of the sacred and mysterious lake of Thangabe, revered by Thonglengpas.

    Neeru Nanda 'The Mahabharat in Himalayan Folk lore' in *The Mahabharat in the tribal and folk traditions of India* (ed) K.S. Singh Indian Institute of Advance Studies Shimla. This article, while focussing on folk lore and folk performances, also mentions mystical and sacred spaces associated with the Pandavas and Kauravas in Uttarakhand.

    Neeru Nanda 'In Search of the Rong' in *Sundaram, Special Issue, the Himalayan Heritage-I,* Vol. No. II, 12 North Central Zone Cultural Centre, 14 CSP Singh Marg, Allahabad. While exploring cross cultural affinities across the Himalayas the article also relates the common strands in worship of sacred mountains and lakes across the Himalayan region into Tibet and the greatest Mountain God of all, Mount Kailash, the Abode of Shiva.

PART III

THE PEOPLE'S FORESTS

The banj forests are very centres of nature's economic cycle on the southern slope of the Himalaya. To destroy them is to cut out the heart and thus bring death to the whole structure.

It is no good putting all the blame on the villagers for the overlopping. In every form of human society, where the population figure is high, and economic stress is felt by the people, things of value have to be protected. The villagers themselves realise fully the immense importance of these banj forests, without which their cattle would starve to death, the springs would dry up and flood waters from the upper mountain slopes would devastate their precious terraced fields in the valleys. Indeed all these misfortunes are already making their appearance on a wide scale. Yet each individual villager cannot resist lopping the banj trees in the unprotected Government forests. "If I do not lop these trees someone else will, so why not lop them, and lop them as much as possible before the next comer." And each villager when you talk to him on the subject, will say "Oh! It is very bad! These precious forests are getting lopped to death. Nobody listens, what is to be done?" And he has as, likely as not, himself been lopping the heads off half a dozen young banj trees that very morning. "That question of his, "What is to be done?" is the all important one. The problem is not beyond solution for if trees are lopped methodically, they can still give a large quantity of fodder and yet not become weak and scraggy. At the same time, if the intruding chir pines are pushed back to their correct altitude and the banj forests are resuscitated, the burden on the present trees will, year by year decrease and precious fodder for the cattle will actually become more plentiful. But all this means winning the trust and confidence of the local people, for the forest department cannot, by itself, save the situation. Nor can it easily win the trust of the villagers because relations.. are very strained....

With the aid of local constructive workers it should become possible to organise village communities and village guards to function along with the forest department field staff which should be increased and also given special training in a new outlook towards the peasantry. In this way it should be feasible to carry out a well balanced long term project for controlled lopping and gradual return of the banj forests.

Such a programme, as suggested above, would not eliminate the revenue obtained from the chir timber and resin but only keep it to its correct proportions. The chir pine is at its best in its legitimate areas which lie between 3,500 feet to 5,500 feet and I may add that if the chir pine forests were better managed they would produce much better results then they do at present. The banj forest are very centres of nature's economic cycle on the southern slope of the Himalaya. To destroy them is to cut out the heart and thus bring death to the whole structure.

Meera Behn

[The Hindustan Times, 5 June 1950;
Meera Behn: A Centenary Tribute;
The INTACH Environmental Series].

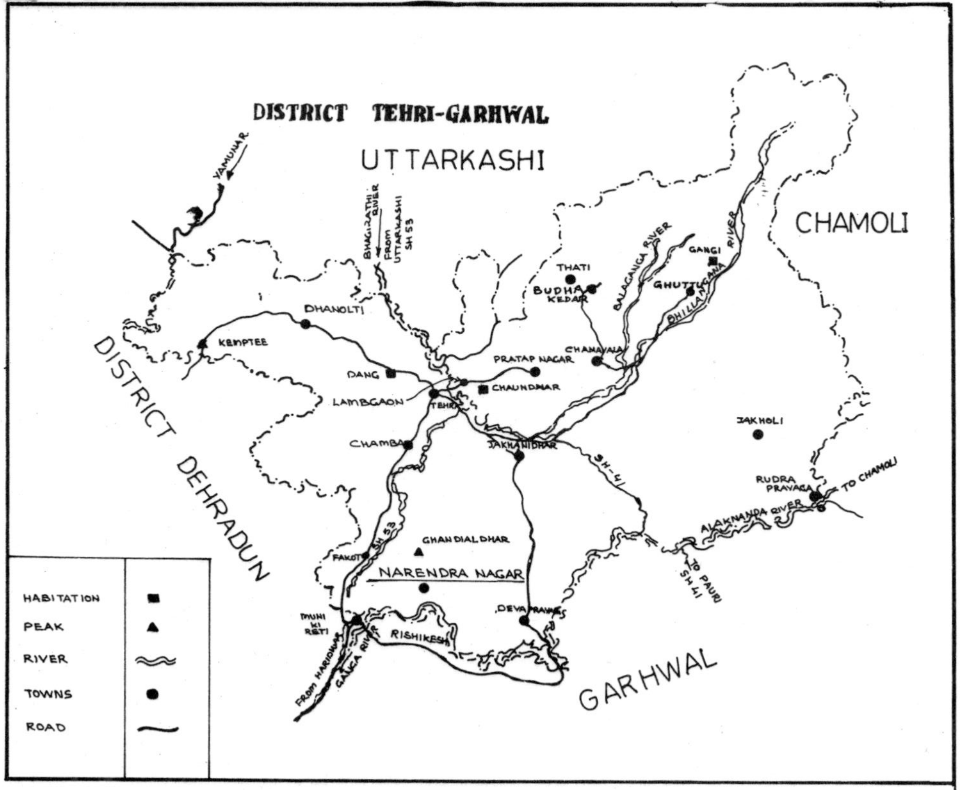
DISTRICT TEHRI-GARHWAL
UTTARKASHI
CHAMOLI
DISTRICT DEHRADUN
GARHWAL
YAMUNAR
BHAGIRATHI RIVER
FROM UTTARKASHI SH 53
DHANOLTI
KEMPTEE
DANG
LAMBGAON
TEHRI
CHAUNDHAR
PRATAP NAGAR
CHANAVALA
THATI
BUDHA KEDAR
BALAGANGA RIVER
GANGI
GHUTTU
BHILLANGANA RIVER
JAKHOLI
CHAMBA
JAKHANIDHAR
SH-41
RUDRA PRAYAGA
TO CHAMOLI
ALAKNANDA RIVER
TO PAURI SH 41
GHANDIALDHAR
NARENDRA NAGAR
FAKOT
SH 53
DEVA PRAYAG
MUNI KI RETI
RISHIKESH
GANGA RIVER
FROM HARIDWAR
HABITATION
PEAK
RIVER
TOWNS
ROAD

CHAPTER 11

People's Protected Forests

PROTECTION VS PLANTATION

To outsiders, the Himalaya are just beautiful mountains. But before the onset of the British Raj, they consisted of well defined territories within which each village exercised its rights of grazing, hunting, and extraction of timber, fuel, herbs, grass, medicinal plants, slates and thatching grass. These rights were accompanied by strict responsibilities for protection and management of natural resources which were regarded as the exclusive property of dispersed communities. To quote Atkinsons Himalayan Gazetteer :

> Under the village tenure which we found obtaining when we took possession of the hills, each village had a certain defined boundary, extending in many instances for miles and miles into dense jungles and to the tops of ridges. These boundaries have never been altered by us... though government is the absolute lord of the soil, the people adhered tenaciously to these old boundaries and look upon any attempt to abridge them as an interference with their rights.[1]

Information pertaining to the remote border district of Tawang in Arunachal Pradesh has been presented by the writer in *Tawang— the land of Mon.*[2] Traditional boundaries, interlocking agropastoral systems and community management of natural resources still existed relatively undisturbed during the authors tenure as a district official in Tawang (1974-77). Hence the information provides a flashback in time, giving an idea of management systems which could have prevailed in the Central Himalaya before they were destroyed with the onset of British rule.

Soon after the British took over, the famous assi sal settlement was conducted in 1823 by G.W. Trail, Commissioner of Kumaon. He recorded traditional boundaries of Garhwali and Kumaoni villages, cautiously stipulating that the boundaries noted by him were in no way an indication of villagers rights to the forest areas since the British Government laid claim to these from the outset as vesting with the State.[3]

Almost a hundred years later, in 1916, A.E. Osmaston, Settlement Officer in North Garhwal, took into consideration all such traditional boundaries while demarcating areas in which villagers were allowed certain concessions for grazing and lopping within notified reserves forests. Osmaston noted that even though the villagers could, in principle, exercise these rights and concessions within the entire settlement block, they preferred to confine themselves within their own traditional boundaries as mutually recognised by adjoining villages. In his words;

> Previous to the forest settlement the whole area was divided amongst neighbouring villages whose boundaries had been defined in the revenue settlement of 1823 made by Mr. Traill...village boundaries are usually well-known amongst the villagers themselves though it would be difficult to define them with any accuracy from existing records and maps and villages will doubtless continue to observe their old village boundaries where it is to their mutual interest to do so.[4]

Atkinsons gazetteer also records the existence of huge rich deodar belts which were khat forests, belonging to various khats, an organisational structure that characterised Jaunsar Bawar, each khat being represented by the village headman or sayana. These were expropriated by the British government as early as 1893, and though a great outcry was raised, it could not be sustained, unlike the subsequent agitation in Kumaon.[5] This agitation took place after notification of reserve forests in Kumaon and Garhwal, of which 400,000 ha were notified in British Garhwal in 1916-17 alone.

As a result of the agitation of 1916-18 and subsequent recommendations of the Kumaon Grievances Committee 1922, a new category of forests came into being.[6] But even while demarcating the new boundaries of Class I and II reserve forests, villagers' ancient ownership patterns and community rights to forests were recorded. Over a hundred years after the assi sal settlement, the British sarkar had to perforce accept the veracity of traditional village boundaries. G.O.No.8295/XIV-285 dated October 22, 1925, of forest department, clearly stipulated that even in the notified reserves, henceforth termed as Class II forests, lopping rights for fuel were granted to :

(a) right holders in that forest where their rights are recorded; and
(b) residents within whose 1823 boundary (Traill's boundary) the forest lies.

The same stipulation covered Class I reserves which were taken out of the jurisdiction of the forest department and placed with the deputy commissioner.[7]

Peoples protection in oak forests of Tehri Garhwal

Interesting strategies for protection of forest-based resources appear to have evolved in Tehri, mostly in reserve forests, even though traditional village boundaries were never documented in the native Tehri State on the lines of the assi sal settlement of Traill in British Garhwal. These have perhaps survived only on the basis of tradition and not always in the remotest areas.

Lamkoti near Chamba, many villages in Saklana area, Jakh. Ramoli, Bheluntha, Chaudhar, Bhenga and Kaulgarh (in Pratapnagar block), Bonga, Thati Kathoor, and others in Balganga valley, Sankri near Kamand in Thauldhar block and a whole cluster of villages around Anjanisein in Jakhanidhar block are only some examples of small communities successfully protecting oak woods and other forests by their own efforts, both in the civil soyam as well as reserve forests of Tehri district.

The oak forests have tended to disappear wherever the community will to guard and preserve their oak forests has been lost and destroyed by forces of urbanisation and change. It can safely be said that almost all surviving oak woods in Tehri are being actively protected and nurtured by village communities regardless of their location either in civil soyam or reserved forests tracts.

Where oak forests are located near settlements (say within a radius of ten km), oak leaf fodder provides the mainstay for cattle. The branches provide excellent fuelwood since oak is a good coppicer and its wood has high calorific value. The leaf compost is prized by farmers. Oak trees are also required to be felled for manufacture of nasooras or ploughs. Besides these multiple uses, oak forests have always been the best guardians of water sources of the Himalayas. This was dramatically illustrated in Dhung Mandhar where there was acute scarcity of water during the drought year of 1987. Despite the overall water scarcity in the area, water sources had some how survived in Lhasi village, because the oak forest from where the spring emanated was being protected by a women's group led by chipko activists. In Ghati village in Pratapnagar block, a small two hectare oak forest is being protcted, without any formal watch and ward, since the past two hundred years. The villagers of Ghati have found that deterioration of the oak forests immediately leads to disappearance of their water sources, hence, for them, protection of this oak forest is a matter of sheer survival.[8]

Very occasionally, inter village disputes come as a boon to an oak forest under lopping stress, provided both parties are equally strong and can manage to mutually prevent each other from using the forest. Some stunted patches of oak around Ghuttu are now regaining strength for this reason. However, oak woods in Ghuttu have generally suffered due to revenue disputes. When one party is in uncertain but adverse possession, little care is bestowed towards future upkeep. Trees are often destroyed by one of the contending parties. Revenue disputes drag on. Though revenue officials are mostly Garhwalis and are well aware of the importance of oak forests, the revenue administration as a whole has no orientation towards environmental protection. Since many oak forests survive on civil soyam land over which the forest department has no jurisdiction, it becomes difficult to work for their protection and care once the land falls prey to litigation.

A surviving oak forest is a sure indicator of a surviving and healthy village community. But though some good oak forests do manage to survive in the remote uplands where no such voluntary custodial mechanisms exist, these high altitude oak forests invariably are under stress since they are used by a large number of villages or village groups (called pattis under native rule).

The oak forests of Ghandialdhar (overlooking the Hyunl valley) and Deodhang forest block (enroute from Budhakedhar to Hathkunidhar) are typical examples of forests which are under pressure but have little hope of being covered by voluntary eco-protection, since eight to ten villages have usage rights in them and it is difficult for these villages to jointly devise norms for regulated exploitation of the biomass. Only remote areas like the Gangi forest reserves retain their pristine glory because the man-land ratio is favourable. But even in the Gangi region, nearer habitation, the surviving oak forests are all covered by some active protection mechanism devised by the community.

Even if diverse human settlements managed to unite and close these remote forests, where would the women forage for fodder and fuel ? *It is a hard ground truth that efforts to encourage protection of oak forests within vicinity of habitation also inevitably increase the lopping pressure on remote upland oak belts.* Since these belong to several groups of villages, no controls operate and for the present atleast their condition is far better than forests adjoining villages.

The case of the Kunjapuri forests, adjacent to Agrakhal, on the national highway to Tehri, illustrates this problem very well. These forests are dying a gradual, albeit painless, death because, despite a strong community will to preserve the trees, the villagers have no other oak forest where burden of extraction can be shifted. Hence, despite the best of intentions and careful lopping of lateral branches, these oakwoods can be seen inching each year towards a slow, inevitable death. The trees would rapidly regain their canopy with only three years complete rest, and the communities using them would gladly close the area, but in such cases the onus of providing alternate fuel and energy sources would rest with the state and the urbanised, privileged sections who have, since the last century, reaped profit and comfort from the Himalayan forests. Since effective provision of alternative fuel and fodder is highly unlikely to take place, the hapless villagers have to manage from whatever has been left to them, so to speak, after the holocaust. Meanwhile the oak woods slowly inch towards ultimate extinction.

The different examples narrated here should suffice to indicate the crucial role played by the oak forests in sustaining the water balance, the gravity of the situation which has emerged after their complete neglect, and the need to immediately put them under maximum protection and care. While the Chipko Andolan had its successes in terms of the ban on commercial extraction, somehow the plantation policies of the state could not be modified to the required extent. Impassioned appeals made by Meera Behn as early as the 1950s also went in vain. Meera Behn herself had tried hard to work through government, first in Pashulok Ashram, in the foothills of Rishikesh, and later in Gopal Ashram in the Bhilangana valley, largely for propogation of local cattle breeds and (in Gopal Ashram) for a people's programme of development based mainly on forest protection and forest management. She had to withdraw from both ventures in despair because of persistent red-tape, lack of decentralisation and highly bureaucratic attitudes of the government staff.[9] This quiet tale of Gandhian discipline and service to the nation, and the failure of her efforts, has been largely ignored, or perhaps just forgotten. But it speaks volumes about the handicap which the nation inherited by virtue of the bureaucracy having so smoothly taken over the mantle of the British Raj in 1948.

Till today, in official circles, there is an absence of policy emphasis and concern for the fast dwindling wealth of the oak forests and other broadleaved trees of the Himalayas, though the entire stress of voluntary protection in the Himalayas is on preserving the oak belts.

Whilst ecologically and environmentally oak must be restored to the hills, it is also important to realise how difficult it would be to regenerate this species unless we are able to obtain the cooperation of the people in protecting the oak forest. This fact was well recognised by the native state and the working plans of Tehri accordingly carried plans of an Oak Lopping

Working Circle, duly divided into compartments for purposes of rotation and rest, but the peculiar nature of the job, which would require very close ground level work with village women, mitigated against its actual implementation.

Any exercise for protection of oak forests, on the pattern of people's protection, would require a great deal of ground work at the grass roots, mainly with women. But our machinery is more used to hieararchical ways of functioning and mechanical fulfilment of targets involving expenditure of large sums of money. Give a state department a scheme which requires management and organisational skills in working with people, and no financial targets to speak of, and they will be completely at sea, if not indignant. Hence, attitudinal changes to enable decentralisation and micro-level planning are also required. Such skills appear to have been developed to an appreciable extent by the West Bengal Forest department and there is no reason why dedicated officers of the department in the U.P. hills cannot profit from the West Bengal experience. The leadership and the impulse, however, would have to come from within the department. Till this happens, world bank schemes for JFM notwithstanding, the contribution of the department towards achieving people's participation for restoring the oak forests of the U.P. Himalaya is likely to remain at the present level of minimal or no intervention.

Protection of species other than oak

Besides oak forests, all kinds of other species are also protected by village communities. Some example are given below :

Chir forests: Contrary to the general conception, (for which some writings of chipko leaders are also responsible), chir forests in the chir zone are valued by villagers for timber and are being extensively protected in civil soyam land as also in the reserves. Examples are Sankri (Thauldhar Block), Bheluntha (Pratapnagar Block), many villages of Pratapnagar Block on the left bank of the Jalkurgad, Kandilgaon near New Tehri Township, Adwani in the Hynul Valley. Further examples will be given in the chapters that follow.

Miscellaneous or Kokat forests: These are being protected mostly in civil soyam land as in the case of Gaujyana, above Patagli (Bhilangana Block). Significantly, this patch of forest is protecting water sources that feed Ghunti shops on the Tehri Ghamsali road. This source did not dry up during the 1987 drought despite acute adversity of climate and large scale drying up of water sources elsewhere. Odada is protecting a good sal forest just above Hyunl river which can be seen across the Rishikesh-Tehri highway near Fakot.

Shrubs and middle storey trees: Tungala (*Rhus Parviflora*), karonda (*Carrissa Sp.*), woodfordia, etc. are being protected as, for instance, by the village women of Motana in Pratapnagar Block. Three separate plots in Motana provide sufficient fuel-wood which is harvested on a three year rotation, two plots being closed while one is kept open.

Survey of people's protected forests in Tehri Garhwal

District officials are generally ignorant of the existence of vast tracts of people's protected forests in the reserves, even though these can be quite extensive. The forest protection committees of West Bengal, though managed by the people, were set up under the supervision and patronage of the forest department and therefore are not strictly comparable with these protection measures which exist independently of government.

Studies on Organisation and Management of Rural Development in Nepal conducted by ICIMOD have, not surprisingly, yielded information which, while retaining the uniqueness of the Nepal heritage and historical experience, is strikingly similiar to the authors findings in Tehri and accounts of peoples protected forests elsewhere in the U.P. Himalaya as, for instance, in Chamoli district in the Alakananda valley and Uphrenkhal in the Dudhatoli forest tracts of Pauri Garhwal.[10]

In a study of the management of forest resources in Sindhu Palanchok district of Nepal, Fisher et al found that indigenous management systems were far more common than expected at the inception of the project, and could be identified, in one form or the other, almost all over the district.[11] In their study (which was coincidentally also conducted in 1987), Fisher et al found that, in Nepal Himalayas, in parallel with the system of legal tenure, there is a local system of claims to user rights and recognition of user rights of other groups, which is perhaps best described as extra legal. It in effect constitutes a parallel indigenous system of tenure. Fisher et al preferred to use the term indigenous to traditional, since the latter implies continuity whereas an indigenous system may be a new development. Their findings agree with the conclusions of different research studies conducted in India, that the focus for peoples protection has to be the settlement and not the dejure panchayat which has become a virtual political body. In their study area of 8 such panchayats Fisher et al could identify 14 examples of people's protected forests.

So far, widespread voluntary practices existing without any external impetus, have come to light in India in the U.P. Himalayas in the wake of the chipko movement, in Chamoli, Pauri Garhwal and later as a result of surveys conducted by the writer in Tehri in 1987.[12]

A systematic documentation of all the village communities which are protecting some patch of forest adjacent to their village area has not been undertaken for the eight hill districts of Uttar Pradesh. Despite the great publicity achieved for Garhwal by the Chipko movement, what has come to public notice is simply the tip of the iceberg. The Chipko movement did not spring up suddenly in the 1970s nor was it the brain child of any single leader. The movement has its deep roots in an abiding tradition which the Garhwal villagers have been struggling to retain, in the face of increasing odds, since the turn of this century.

It would have been impossible in the short period of leave available to the author in 1987 to survey all the 1983 odd villages of Tehri district in order to identify all those which are voluntarily protecting forests. However two different but fairly extensive surveys could be completed over a ten year period of continued association with the district administration.

In 1987, with the help of Deputy Conservator of Forests of

Tehri Dam I Shri S.T.S. Lepcha, a considerable amount of data could be collected with reference to certain Ranges and Forest Guard beats in the Bhilangana catchment. Almost ten years later, another survey was conducted through the women workers of Mahila Samakhya, an organisation for women's empowerment which was just being set up in 1987 by the Ministry of Human Resource Development.

The results of the survey conducted by the forest department in 1987 are analysed and presented here, while the information gathered through the good offices of Mahila Samakhya is analysed, with reference to eco-management practices, in the chapters that follow.

TABLE 1

Number of Villages Voluntarily Protecting Forests in Vicinity/ Adjacent to the Village (Area in Ha)

Range/ Forest Guard Beat	*No. of Villages*	*Area Protected Civil Soyam*	*Area Protected Reserve*	*Total Area*	*Average Area per Village*
Balganga	26	110	1021	1131	43.5
Dharkot	6	-	392	392	65.0
Dhung	10	64	285	349	34.9
Total	42	174	698	1872	44.5

The data indicates that over 90 per cent of the protected area falls in the reserve forests, clearly reinforcing one of the main arguments of this study *viz.* that the major burden of biotic pressure rests on the reserve forests, and that it is these forests, (as against civil soyam areas), which need to be the focus of protection and afforestation. It is indeed remarkable that, in their own humble way, villagers are uniting to protect (at considerable sacrifice to themselves) reserve forests where their legal rights are far more circumscribed than in civil soyam. Yet they prefer to protect these because soil moisture and other conditions conducive to successful afforestation are more favourable in the reserves, as against the soyam area, and because their dependence on reserves is greater. The protection systems are also perhaps a remnant of the age-old rights and responsibilities which their forefathers might have once exercised in these forests before the onset of European systems of silvicultural management leading to expropriation by the state.

It is indeed a sad commentary on bureaucratic diagnosis and treatment for the Himalayas that, over the past twelve years, 314 sq. km. of land was taken up for plantation in the civil soyam areas of Tehri Garhwal, as against only 35 sq. km. in the reserve forests. It is astonishing that villagers on the other hand have been doing exactly the opposite. Obviously the common man living in the midst of a fast degrading environment knows both where the problem lies and how to tackle it in the cheapest and most efficient manner.

The average size of plots protected is about 45 ha but varies from 35 ha to 65 ha, indicating the need for ground level micro-planning for each and every group of villages in a catchment.

Though it was difficult to obtain accurate figures of the number of villages falling in each plantation division, not to speak of the ranges and forest guard beats, from general knowledge and discussions with the field staff the following estimates were arrived at :

TABLE 2

Proportion of Villages Protecting Forests

Area	*Total No. of Villages*	*Protecting Forests*	*Percent*
Balgaṅga Range	80	26	33
Dharkot Range	35	6	17
Forest Guard Dhung	25	10	40

It can be seen from the above that the practice of eco-restoration through simple closure and protection is fairly wide spread and if encouraged and recognised by Government, more villages might come forward to actively protect their adjacent forests in this manner.

TABLE 3

Village Protection by Species

Area	*No. of Villages*	*Banj (Oak)*	*Chir*	*Chir Oak Mix*	*Kokat*
Balganga	26	3	4	19	-
Dharkot	6	4	-	-	2
Range Dhung Beat	10	8	-	-	2
Total	42	15	4	19	4

The data indicates that over 80 per cent of the protected species comprise oak, clearly reinforcing one of the main arguments of this study *viz.* that the major burden of biotic pressure rests on the oak belts in reserve forests, and that it is these forests, which need to be the focus of protection and afforestation.

TABLE 4

Village Protection by Size of Plots

(Area in Ha)

Range	*Below 20*	*20–25*	*30–35*	*40–50*	*60–70*	*100*	*Above 200*
Balganga Range	2	4	9	7	2	1	1
Dharkot Range	2	2	1	-	-	-	1
Dhung F.G. beat	2	4	1	2	-	1	-
Total	6	10	11	9	2	2	2

Table 1 has shown that the average plot size was around 45 ha, whereas from the above figures it is seen that about two thirds of the plots are actually 35 ha or less. It is apparent that a village cannot normally afford to effectively protect very large sized plots. Policies of the plantation divisions, therefore, have also been on the right lines since the department seldom selects plantation plots exceeding 25 ha even though, in terms of time and supervision, this strategy has put a great strain on the forest department.

TABLE 5

Duration of Protection by Villagers

Area	*Total No. of Villages*	*Since 50 years*	*Since 30 years*	*Since 20 years*	*Since 10 years*	*Lessthan 10 years*
Balganga	26	7	10	9	-	-
Dharkot	6	1 -	-	5	-	
Dhung F.G. beat	10	-	-	7	-	3
Total	42	8	10	16	5	3

The above table brings home some telling facts. Over seventy five per cent of the villages have been guarding and protecting their forests for more than twenty years (no less than twenty per cent have actually been doing so since the past half century). Exactly twenty per cent of villages, on the other hand, have been protecting their forests since ten years or even less. This indicates that, given the appropriate set of circumstances, ancient village traditions of eco-management may still be valid and can be easily revived, since the identification or sense of 'belonging' with forests in the vicinity of human settlements still survives in embryonic form in almost every highland village in the Himalayas.

In the native state of Tehri Garhwal the State Forest Department had set up deodar and blue pine plantations in what is now the Dhanolti Kanatal belt. Of small patches elsewhere, Pratapnagar is a notable example. All these plantation areas were tended and cared for over a fifteen to twenty years time span. The plantation areas were passed down, lovingly as it were, from each DFO and each Working Plan Officer to the next. Hence, in the Working Plan of 1939-40, P.N. Khosla could record that

> all deodar plantations are coming up very well, are receiving all the care and tending required and in course of years are bound to prove useful and remunerative.

The Forest Department had traditionally managed plantation areas over a time-span ranging upto 50 years, because that is the time it takes for full restocking of a degraded forest or for bringing up a new plantation. Thus the Forest Department strategies for protection and nurture are no different from traditional eco-management practices of people's protected forests in Garhwal, even though the department concentrates on mono-culture and the people on regeneration of natural forests.

The new policies of the National Wasteland Board and the new concepts of Social Forestry of the decade of the seventies and eighties, however, just assumed that department plantations needed no after care. The compulsions of the twenty point programme and the political commitment for planting several lacs of trees every year, coupled with the absence of after-care, and protection — all these policies, in the most callous manner, made a mockery of plantation works of the Forest Department in Tehri Garhwal.

The Costs of Protection versus Plantation

What are the hidden costs of protecting some areas as against others? In a few cases, the villagers had to sell off their surplus cattle, but in most cases biotic pressure simply shifted elsewhere. The classic case of Bonga village, in the Balganga valley, dramatically illustrates the basic predicament posed by village protection efforts. Just above the last line of sound healthy shoots of the oak forest in civil soyam land protected by the villagers of Bonga, there lies a vast patch of degraded oak scrub of the adjoining reserve forest. Since the daily needs of the village had to be met in any case, biotic pressure simply shifted to the trees which were located adjacent to the protected area. To their utter dismay, the Bonga villagers found a second bald patch appearing in the forest above their village, even as the earlier arid stretch got time to recuperate. Unable to control their women, desperate villagers finally took their village devata (Temple Deity) around in a parikrama (circumambulation) of the area where biotic stress had shifted, thus magically 'sealing' the boundaries. No village woman dared to enter that plot thereafter and their drudgery only increased, because now they had to trudge further afield for their daily headload of fuel and fodder.

In many villages biotic stress simply shifted to more interior and distant areas which belong, either to the village itself, or, more often, to a group of villages in common. However we do find instances such as Motana (Pratapnagar Block) where the biomass generated in the village commons is sufficient to enable the village women to follow a system of rotation and rest for extraction of biomass.

What is the cost being incurred by the people in terms of actual expenditure involved for policing and protection ?

Separate data could not be gathered regarding salaries being paid to village chowkidars. However, it was common knowledge that the going rate was around Rs.500 p.m. in 1987. This would amount to annual expenditure of Rs.6,000 per village. Assuming for a moment that each village spent this amount per annum, whether in cash or kind, we can conclude that approximately 1900 ha of excellent restoration work in the Himalayas covered by this survey alone, was being undertaken by villagers themselves at an annual recurring cost of Rs.2,52,000 per annum. This roughly amounts to an average cost of Rs.133 per ha. Taking a ten year period as the norm for attaining a fair degree of density and restocking of the forest, this gives a decadal afforestation cost of Rs.1,330 only per ha.

This cost of afforestation compares very favourably with the cost of plantations raised by the plantation divisions of the Forest Department. The expenditure norm as per schedule of rates approved by the Conservator Bhagirathi Circle in 1987 was Rs.4,000 per ha (exclusive of establishment costs). Within this norm 1600 trees were being planted per hectare. Survival rates ranged from 5 per cent in the worst areas to 80 per cent in the best. On an average, in the best of years, the Forest Department reported survival of sixty to sixty five per cent for the first year. Given the present arid conditions and inappropriate strategies, at an optimistic estimate, the average survival of three year old plantations in the Garhwal Himalayas fluctuated between thirty to fifty per cent. These plantations were also given up after three years when the part time chowkidars, (with monthly salaries ranging from Rs.75 p.m. to Rs.175 p.m.) were dismissed.

Thus, even while assuming a 50 per cent survival rate, the cost of raising one hectare of successful plantations by Government under the present strategy is much higher than Rs.4,000. Taking the cost of each tree @ Rs.1.50, the notional loss of trees that do not survive is about Rs.1,200 per hectare. This is the minimum loss. In adverse soil and moisture conditions, survivals can drop even to 20 - 25 per cent.

In the village protected forests, the question of mortality of natural rootstock and natural shrubbery does not arise and chowkidars are a permanent feature of the management system. Infact, studies have shown that protected natural forests regenerate better and yield greater quantities of biomass and minor forest produce than areas under commercial monoculture plantations. The economics of plantation *vs* protection and the cost-effectiveness of the latter have been demonstratively established by the Arabari experiments in West Bengal. They have also proved that regeneration strategies, while being cheaper, actually bring in greater income for the poor than plantations and silvicultural management for sale of pole crop and timber, largely because of the proliferation of under-storey and middle-storey trees, shrubs, creepers and grasses which takes place under the former system. Studies have indicated that an active woman could obtain an income of Rs. 7 to Rs 10 a day from one hectare of mixed sal forest after a rest period of three to four years. By contrast the system of sharing 25% income from revenue on sale of sal poles would only yield Rs 1.4 per hectare daily income if averaged over a ten year period. Over and above this is the reliability, continuity and sustainability of minor forest produce as a source of income in the sal forest tracts of West Bengal, Bihar and Orissa.[13]

By any standards of cost benefit analysis, an investment of Rs.133 per ha annum for successful restoration of forest cover is abysmally low. This compares well with other research studies conducted in various parts of the country which have established that the cost of joint forestry management based on protection ranges from Rs. 0-500 per hectare as against plantation costs of Rs. 5000–10,000 per hectare.[14]

In the past decade the Civil Soyam Plantation Division alone spent Rs.2,83,00,000 for plantation of about 11865 ha of land. At a decadal investment rate of Rs.1,330 per ha for protection by well paid village chowkidars, this budget allocation could have helped to effectively and continuously protect 213 sq km of land over the same period. It is sad but pertinent to note that all the plantation divisions of the Bhagirathi Circle combined could not cover more than 318 sq. km. for afforestation and plantation in their entire life span upto 1987.

The advantages of this strategy for the Himalayas have already been spelt out. The Himalayan region has not been considered by some researchers as an area which offers high potential for yielding significant results through protection for regeneration and this is no doubt true if the pace of recuperation is compared with the much quicker pace in sal forest tracts.[15] If completely protected for ten years, however, the forests have been found to improve to a significant extent, both by way of density and biomass in the middle and understorey. Because of this visible improvement and returns by way of grass, leaves for fodder and reappearance of water sources, the motivation for continued protection by villagers, as also their total participation, could be banked upon if the state decides to adopt strategies of protection, as against plantation, for restoring the Himalayan forests.

What are the lessons for Government ?

Out of 2,000 ha of village protected forests covered by the survey described in this chapter, almost 1,800 ha were being protected by villagers in reserve forests adjoining their village. *This shows that biomass needs can be met by a combination of revenue commons and reserve forests and that the divisions between them are purely paper divisions.* This fact has major and far reaching implications for future investment patterns in plantation, protection and restoration as also for restructuring the management set-up of the forest department.

The forest department would have to acknowledge that reserves can and are being managed by the people who are playing a key role in their protection. While this tradition is over 50 years old, it is still alive and vibrant, as there are a number of new entrants.With such a widespread network of protected village forests, many of which were raised from scrub simply by affording complete rest and protection, the future of the Himalaya does not seem so grim.

A new strategy could be adopted, wherein the bulk of government investment would be spent on salaries of chowkidars to be appointed and paid by village groups. But the strategy would have to take care not to destroy the self-reliance of these small communities and the control they exercise over their chowkidars. It is a sad and punishing fact that no aid, whether from government or external agencies, comes completely without strings. If the department staff starts interfering with the appointment and control of the chowkidars, the scheme would be agonisingly counter-productive. In Nepal, people's watch and ward which existed in a certain panchayat, was discontinued when an externally sponsored plantation project appointed its own chowkidar. However, in this case, consensus for protecting the forest remained.[16] Hence, interventions, if at all, would have to be very carefully planned and worked out. Unfortunately, the van panchayat rules also stipulate that the panchayats are to appoint staff only with the prior approval of government.

Lump sum grants (without stipulating that the money be used for watch and ward), as incentives to village panchayats which are voluntarily and successfully protecting forests might lead to a different kind of problem. Protected areas are seldom coterminous with panchayat boundaries. Many panchayats have disputes between settlement groups within the panchayat area and money coming in to the panchayat as a whole for areas being protected by only a segment would tend to defeat the purpose.

User and protector groups for forest management operate on the basis of consensus rather than conflict, and are quite distinct from elected panchayats whose management is generally confined to a single neta (leader) and his coterie. Hence, when the village scenario is seen in its entirety we seem to have before us a situation where good intentions would not suffice unless the interventions are very carefully designed by staff and voluntary workers as, for instance, the Mahila Samakhya grass-roots workers working mainly with women.

In an optimistic scenario, such innovative strategies could provide maximum employment opportunities and security on a sustained basis for poor families, while also creating a network of motivated village guards who would supplement the department efforts for policing and providing fire protection to the reserve forests.

The last two chapters of this book document eco-management practices being followed by villagers for preserving and regenerating their community resource banks in Pratapnagar, Bhilangana and Jakholi blocks. These facts, gathered by the Mahila Samakhya village workers, could be profitably studied by forest department officials while designing and mapping out interventions for restoring the Himalayan forests through the latest initiative of Joint Forestry Management which is just entering the scene, with world bank sponsorship, in Uttarakhand.

NOTES

1. Atkinson, *The Himalayan Gazetteer*, Volume III Part I, p. 315, 'Waste for grazing.'
   Also see Ramachandra Guha, *The Unquiet Woods*; Oxford University Press; 1989; pp. 28-34, "Forests and social institutions in the indigenous system". Guha quotes Lucien Febvres description of Andorran highlanders in A *Geographical Introduction to History* (London, 1925) and Robert Netting, *Balancing on an Alp* (Cambridge, 1981), to suggest that, in its democratic characteristics and traditional patterns of natural resource use, Uttarakhand is representative of mountain societies in general.
2. Neeru Nanda, *Tawang : The Land of Mon*, Vikas Publishers, New Delhi, 1982.
3. A.E. Osmaston, para 19, p.8.
4. *Ibid.* Also see *The Himalayan Gazetteer*, Vol. III Part I, p.295, Mr. Batten is quoted thus;
   "Large portions of wastelands including whole ranges and their vast forests, have been included from olden time in the boundaries of

adjacent villages.... such a division has been found useful in giving separate tracts for pasture for the cattle of different villages... I therefore take this opportunity of asserting that the right of government to all the forests and wastelands.. remains utterly unaffected by the inclusion of certain tracts within the boundaries of villages".

Also see p.315.

5. *Ibid.*
6. Ajay Singh Rawat, *Garhwal Himalayas : A Historical Survey 1815 - 1947,* pp.62-63.
7. *Ibid.*
8. For an excellent research study on the connection between broad leaf forests and water supply see A.N. Chaturvedi; 'Environmental Value of a Forest in Almora,' Anil Agarwal (edited); *The Price of Forests* CSE, New Delhi, 1992. This article clearly establishes the value of broad leaved forests in the catchement of Almora town for supply of water at a capatalised value of Rs. 60 lakhs, the value of the forest cover per hectare for water supply alone amounting to Rs. 31,630.
9. Mira Behn; (Madeleine Slade;) *The Spirit's Pilgrimage;* Longmans, Great Britain; 1960; ch.LXIX-LXXI; pp.299-308.
10. R.J. Fisher et al *Management of Forest Resources in Rural Development a case study of Sindhu Palchok and Kabhre Palanchok district of Nepal ICIMOD* Joint Case Study (unpublished).
11. Protected Forests of Chamoli district have been documented only in the case of forests planted and protected by the Dasholi Gramswarajya sangha founded by Chandi Prasad Bhatt of Chipko fame. In Uphrenkhal a voluntary organisation of local leaders is networking for successful management of the vast grazing areas and forest tracts of the famous broadleaved forests of Dudatoli in Pauri district.
12. Neeru Nanda; 'Local Management of Community Property Resources in the Central Himalaya' in Anil Agarwal (edited) *The Price of Forests.*
13. K.C. Malhotra, Mark Poffenberger (ed) *Forest Regeneration through Community Protection: The West Bengal Experience, West Bengal Forest Department,* 1989.
14. Poffenberger, McGean & Khare; 'Communities Sustaining India's Forests in the Twenty First Century', in *Village Voices Forest Choices;* OUP ; 1996; pp. 54-55.
15. Manjula Bajaj; 'Natural Regeneration Versus Afforestation: An Examination of the Potential for Investment in Natural Regeneration of Degraded Forests with Community Participation' in Anil Agarwal (ed.) *The Price of Forests.*
16. Fisher et al *op. cit.*

CHAPTER 12

Eco-Management Practices in People's Forests

Mandhar gram sabha has been protecting about 50 hectares of forest over the past 60 years. The forest area has been divided into compartments in the best traditions of departmental management, almost on the lines of the Oak Lopping Working Circle, formulated by Working Plan Officers of Tehri Forest Division, but never translated into practice by the forest department. The compartments are opened sequentially and collectively lopped in accordance with schedules laid down, from time to time, by the village assembly.

H.D. Semwal, a forest guard, who used to work closely with village groups, told me of an instance when the Assistant Conservator happened to tour the oak forests of Mandhar on the very day when one of the compartments had been opened by the gram sabha. There was hectic activity in the oak woods. All adults were out in full force, armed with their drantis (sickles) and the forest resounded with the noise of branches being lopped and dragged down the hill slopes. The ACF was momentarily enraged, having jumped to the conclusion that mass deforestation was taking place, right under his nose, with the help of conniving forest guards! He was pleasantly surprised when Semwal explained the whole system of local management to him and convinced him of the value of people's voluntary protection and the unique effectiveness of their management practices. "I tried to make him understand the system", Semwal said. "We are able to see this forest in good condition only because the people are managing it themselves, by appointing and supervising their own paid chowkidars, and by lopping the trees in rotation. Do you think it could have survived otherwise? It is impractical to expect a lone forest guard to successfully safeguard these scattered jungles."

There is much to be learnt from management practices adopted by local people in the forests being preserved by them.

The narratives presented here reveal the inescapable facts that, wherever small communities have attempted to stem the tide of deforestation, their strategies have been based upon fairly sound eco-management practices and have been able to show marked success, as against the overall dismal survival record of government plantations.

Protection of trees

Be it private ownership of trees or ownership of a restricted area by a specific village, exclusive ownership has proved to be the key factor in successful protection efforts at the micro-level. Oak forests or parmong of Tawang in Arunachal Pradesh have survived even the onslaught of extended administration simply because each tree is privately owned and passed on from father to son.[1] Private ownership of trees like poplars and willows is quite common in Ladakh and Kashmir. The Apatani Plateau in the Ziro valley of Arunachal Pradesh is well-known for the stately blue pine forests which surround the valley on all sides. But it is not so well-known that these trees are lovingly planted and reared by Apatani families, being passed down from father to son.[2] Somewhat similar practices could be identified by this writer in Tehri Garhwal in remote Gangi, where Kharshu trees (*Quercas semicarpifolia*) form a significant corner stone of the economy. Kharshu leaves remain green in winter, providing the only available source of fodder for cattle and sheep in winter, when the area is snow bound. Since per capita availability of trees is sufficient to meet local requirements, villagers have been able to divide the trees amongst themselves, usufruct being privatised while retaining community ownership of land. Certain villages in Pauri Garhwal have planted civil soyam lands with trees which are allocated to individual households. In Bheragaon village, in the Bhilangana valley, sailma grass grown in community land has also been similarly divided between families. Such instances of individual ownership, though rare, can be found only in small, remote and cohesive village communities where per capita availability of land is more than adequate and usufruct can be equitably divided. In many parts of Jakholi block, e.g. Gauriya, Manjiyari, Shankari and Sumarth, to name just a few, common lands have been parcelled out amongst the villagers, quite on the pattern of a local tree farming project ! In this way, the trees are being lopped scientifically and are properly protected. However, some families are seen to have enclosed more trees than others and voices against inequitable distribution are now being raised by those who feel relatively deprived.[3]

Basic principles of protection and eco-management of forests

Villagers and local leaders used to repeatedly question me. "Why does Government plant so many trees without appointing full time chowkidars? Let them not plant anything. Just appoint a chowkidar and pay him a living wage to protect the natural shrubbery and rootstock. In ten years it will grow into a beautiful forest without any more effort on your part."

The stories of Lambkoti and Ramolgaon will illustrate this strategy.

In Ramolgaon village, in Pratapnagar block, eco-restoration had become an issue for the entire community because the village oak forest area had been wrested from a neighbouring village after a long-standing dispute, as a result of which the losing party had spitefully razed it to the ground before handing it back. Ramoli villagers had to suffer untold hardship. Not only was fuel and fodder hard to come by, adjacent water sources and springs also dried up. Such was the extent of hardship that the young men found it difficult to obtain brides. No family was willing to send their girls to the village.

All this took place about thirty years ago, when the community decided to protect the forest by appointing a well paid full time chowkidar whose only job would be to police the area and punish transgressors. Grazing of cattle and lopping for fuel and fodder were strictly forbidden. After fifteen to twenty years of watch and ward the trees regained strength. Water sources reappeared. Oak leaf fodder can now be harvested. For marriage feasts, a dead and dying tree is identified by the chowkidar and felled in his presence. This forest is now the pride of Ramlogaon and the chowkidar who first patrolled the area is a noted and respected figure in the village.

Lamkoti in Chamba block (just off the route to New Tehri Township) forms an interesting contrast to Ramoli. The decision to protect its adjacent oak forest was also taken about thirty years ago, mainly because of hardship due to scarcity of fodder and drying up of water sources. However, Lamkoti never had a permanent chowkidar as in Ramolgaon. The householders share this responsibility annually in turn and receive payment mostly in kind. Villagers told this writer that, even though they have a chowkidar, responsibility for apprehending offenders is shared by all. Their forest is very near the village and the sound of an axe carries far over hill and dale. Anybody who happens to hear the sound can move quickly to catch the intruder. Now Lamkoti has a good medium sized forest which has attained this state, as in Ramolgaon and innumerable other villages, by achieving an active consensus for closing it to grazing and excessive lopping and by also ensuring fire protection.

A variety of eco-management practices are adopted by villagers for indefinitely prolonging life of trees in these protected areas.

In totally denuded areas, complete closure of the forest is often resorted to. The forest is opened once or twice a year only for harvesting of grass which can be particularly nutritious in oak forests. If it is not grazed or cut during the rains this grass gives an excellent crop, once in November, and again in March or April, after the winter rains. In areas where the trees can tolerate lopping, entry into the forest with any kind of sharp instrument is normally forbidden, while harvesting of leaves and twigs is permitted. This can be done by breaking lateral branches. A third method of rotational lopping of trees is followed in some villages where a tree which has been lopped once is not touched by others for atleast two years, or till such time as it regains its canopy. This system can only succeed where biotic pressure is limited and the village forest area is also well known and circumscribed. This practice was found to be prevalent in Bheragaon village in Ghuttu area.

In most protected areas entry with sharp weapons is strictly forbidden. Often the village forest guard, who is appointed by the community, proudly shows off his horde of drantis (sickles) and ropes confiscated from offenders. Many villages regulate lopping by opening the forest in a controlled manner so that all women proceed together, with their drantis, along with the forest chowkidar for collection of leaf fodder.

Contrary to the general assumptions in environmentalist circles (as also in the Wasteland Board), the concept of social fencing involves payment of money to chowkidars and is not based solely upon the noble principle of 'enlightened' and 'participative' management by 'motivated' and 'aware' villagers. No doubt the village community has to be enlightened and aware to resolve to protect its adjacent forests but such resolution by itself is not enough. The villagers explained to this writer, that forests have to be protected not only from black sheep within their own village, but also from people of other villages. Many people would be only too ready to intrude and make a quick get away with the spoils. Thus villages lying on the left bank of Jalkur gad *viz.* along the west facing slopes of Pratapnagar block have to maintain a constant vigil to protect their forests against women who come, desperate for fuel, from across the river where the east facing slopes are a pitted and scarred moonscape, as bare and deserted as some alien planet. Hence villagers invariably appoint a full time chowkidar who has to be paid in cash or kind through contributions shared equally by all households. In certain villages in Chamba block the chowkidar's monthly pay exceeded Rs. 500 p.m. in 1987, with an average monthly contribution of Rs. 8.00 - Rs.12.00 being gathered from each household. This is a far cry indeed from the policies of the Wasteland Board in the nineteen eighties when investment in maintenance and protection of areas where trees had been planted at great expense to Government was not catered for. Bureaucrats sitting in Delhi are unable to realise the grim realities of "social fencing", in the U.P. Himalayas — what a struggle it is for even a small village to protect and police its area in situations of such desperation as exist in Tehri Garhwal. Unfortunately, the idea of restoring natures bounty, where it still exists, through simple protection has not yet become quite the fashion in Government circles.

Documentation of forest management practices by Mahila Samakhya

Mahila Samakhya was set up in 1987 as part of the New Education Policy thrust of the Rajiv Gandhi government which sought to link rural literacy with womens empowerment in rural and backward pockets of the country. The organisation in Tehri now has a network of about 200 village groups or sanghas, with ten such groups being looked after by a woman worker or sahyogini who is invariably selected from amongst them.

In the first five years of its existence, the Mahila Samakhya Tehri got intensively involved in womens problems stemming from a fast degrading environment, and these came to the forefront, along with a pre-occupation with the anti-liquor movement. The small, graphic documents brought out by Mahila Samakhya as a result of these exercises gave evidence of their potential for undertaking more serious surveys.[4] After being briefed about the survey undertaken by the author in 1987 (described in the previous chapter) a preliminary documentation of forest resource management by villages extending across four blocks was completed in 1996 by Kusum Rawat, District Project Coordinator, Mahila Samakhya.[5] This had to be followed up with more briefing sessions with the sahyoginis who came up, in October 97, with extensive notes that required a great deal of labour by way of editing and compilation, but proved well worth the trouble, since the information they yielded definitely establishes the almost universal prevalence of people's management practices for protecting community resource banks throughout the district.

Protected Forests in Pratapnagar block

About 42 village groups, some consisting of two or more settlements were identified and their management practices noted by six sahyoginis of Pratapnagar. The statistics given are open to correction, since the extent of the protected area, as also the number of trees and the duration of protection, are based on very general information gathered from villagers. Some of the important cases given below are, however, vital for the light they throw on the whole process of voluntary watch and ward, formulation and enforcement of rules, punishment of offenders, distribution of biomass and conflict resolution. Hence, while the figures of area protected, duration of protection, identification of reserves might be inaccurate in places, the processes explained by the women workers are authentic and illuminating.

Shukri Gramsabha

The seniormost group is that of Shukri gramsabha which has been protecting over 500 hectares of forest, in the civil and reserves, since the past hundred years. A van panchayat was also formed in this village in 1992, but it does not seem to have affected the existing informal management systems. The latter cover the reserve forests as well (in areas such as Maimendya, Ghat and parts of Budagkholi), whereas the van panchayats can only be formed in civil soyam lands, and are consequently hampered from inception in this respect. As in Orissa, the villagers seem to have remained ignorant of the establishment of formal structures such as the van panchayat and the previous informal procedures for management of biomass extraction continue.[6]

The prevalence of traditional boundaries within the reserves emerges clearly here, since parts of Budagkholi in the reserves are under traditional usage of Shukri, while other parts of Budagkholi (in the reserves) are under protection of their rival village, Ramolgaon, whose saga has been narrated above.

The stories regarding the genesis of this protection system, as related by Shukri villagers, indicate that even indigenous arrangements sometimes need to be enthused or reactivated by a catalytic agent. Dedicated care of this vast jungle was started by a village elder called Jagat Ram. He is reputed to have jealously guarded the trees like his own children and his voice could often be heard, hollering after errant women who had perhaps lopped the head off a young banj tree. "Those who lop the heads off the banj of this jungle—be sure you are dealing the same blow on the heads of your children!" His tireless vigilance, coupled with imprecations and curses, obviously bore fruit. The villagers took it upon themselves to appoint a full-time chowkidar who was paid through an annual contribution, commonly known as phant, of 25 paise per household. This fees has now gone upto Rs. 50 per annum per household. Shukri villagers first appointed a woman as chowkidar. Munjari, the first chowkidar, must have been good at her work because, gradually, more areas were handed over to her for protection in Upaphadi, Payandabar and Patha in the civil and Ghat in the reserves. Gradually the area under voluntary watch and ward expanded to its present dimensions.

The villagers appoint the chowkidars and also change them when required. They are given the powers to seize the drantis (sickles) of offenders and scold and disgrace them, but the power to impose fines vests with the panchayat. The chowkidar simply reports the offence, after which the panchayat summons the offender who can be fined Rs. 25 for the first offence and double the amount for the second offence. Repeated thefts and violation of the rules agreed upon by the village assembly can lead to debarment of the whole family from usage of the specific biomass which had been stolen from the forest.

Dang and Ramolgaon

The example set by Shukri was perhaps followed by their close rival group, which has now been protecting about 50 hectares of forest in the reserves (Naulan, Gherka etc) and the civil (Patala, Chamlyanu etc), since the past 50 years. Perhaps because of the acute scarcity conditions which initially prevailed in Ramolgaon, the rules of the system appear to have been very carefully worked out in order to ensure equitable distribution. The banj leaves, twigs, branches as also grass loads brought down from the jungle, are weighed at the entrance, and only equal measures are allowed to be taken out free. The rest are auctioned amongst themselves, but the system is quite fair, because the women themselves buy up the excess, and no woman will take the trouble of bringing down extra load which she cannot later procure. Further, the money raised is given to the chowkidar and adjusted against annual phant or fees. A similiar distribution practice was found prevailing in many villages in Bhilangana valley.

These villagers reportedly inspect their forest before it is opened every year, and if trees are found damaged or felled, the chowkidar is fined by reducing his phant or by changing him.

Pansut Gramsabha

The people of Pansut were actively influenced by Shukri and followed their example, not only by closing their civil lands to indiscriminate exploitation, but also by taking acorns from Shukri banj jungles and planting up their own oak belts, apparently without any help from government. They now have a 100 hectare protected plot of broad leaf forests which they have been guarding since the past 40 years. They had started off by paying the chowkidar with a phant (annual contribution) of Rs. 5 which now stands raised to Rs. 100 per household.

The village assembly meets annually to decide on appointment of the chowkidar and other rules for regulating extraction. The chowkidar is empowered to catch thieves in the forest and rules are laid down with regard to disposal of the drantis and grass etc seized from offenders. While the fines in Shukri are pretty steep, the fines here are Rs. 5, 10 and 15 only for successive offences.

Kharoli Gramsabha

They have been protecting about 40 hectares, since the past 20 years, in the broad leaved jungles of Semaldhar and Chamlyanu, and in the chir forests of Mahatkhal in the reserves. The phant in this village is Rs. 60 per household. In addition, cereal is given to the chowkidar who is appointed annually by the village assembly which meets to pass a resolution regarding the appointment of chowkidar and other rules which will govern the system.

Majaf, Silwalgaon, Ghanti and Kuran

These four settlements have a very large broad leaf forest of banj, burans, ayar etc extending over 5000 hectares. Traditionally, it had been divided into three parts, which used to cater separately to the needs of these groups. But conflicts developed because Ghanti and Kuran women preferred to cut grass in the nearby areas, which fell within the traditional jurisdiction of Majaf, in preference to their own area, which was located at a remote distance. Since daily headloads of fuel and fodder have to be fetched from the forest, conflicts and clashes between the women became a commonplace occurrence and ultimately reached a point where the men just decided to get together and put an end to these quarrels "once and for all". A jungle samiti (forest committee) was formed and rules were drawn up for collective management and controlled exploitation of the biomass. Two chowkidars were appointed to guard the forest which is now opened under strict regulations. Each woman has to deposit her annual phant of Rs. 70 before she can enter. Her loads are also weighed on return and a stipulated amount of 'puli' of grass is kept for the chowkidar. Fine money is deposited with the samiti.

This male dominated samiti is managing the forests quite well in terms of protection. The trees have gained in strength and canopy but the sahyoginis (who are quite aware of their rights), seemed a bit discontented with the lack of participation of women and poor feedback regarding receipt of fine money and its utilisation.

Khet, Dulyan

These villages have their separate grass 'bittas' (plots) which are regularly opened, but, unlike the others, they have no forests of banj in their civil land. Hence they are protecting about 100 nalis of broadleaf forest in the reserves, just above their settlements, in an area called Thankula and Dulyab. An annual phant of Rs. 70 is given by each household towards the pay of the chowkidar who is allowed to retain the fine money realised from offenders. Fines can go upto Rs. 100. Though steep as compared to other villages, payment is ensured because failure to comply leads to effective debarment from the forest—a sanction which is implicit in all villages operating the fines procedure.

These villages are notable for the rotations being practiced. Since they started guarding the forest, just 15 years ago, the leaves and grass in the forest are extracted collectively once a year, while lopping of branches is presently allowed only once in three years.

Mishravan Gaon and Dhargarh

While both villages have land in the civil soyam, which they use mainly for extracting grass and shrubs, they used to have to go quite far for fuelwood and banj leaves, till they decided, 12 years ago, to close the reserve forest above their settlement. The chowkidar is paid through an annual phant of Rs. 60 levied on each household. This area, of about 200 nalis, now has a fair number of standing trees, some of which have been planted by the villagers.

Protection of forest species other than oak

While there was an impression, at the commencement of this study in 1986 that safeguarding procedures are only being adopted for banj jungles, it has emerged over the years that all forest species are being managed for optimum and sustainable use.

Okhala I and II, Gvad I and II,Chathi and Raulakot villages fall in the lower altitude khair and sheesham belts where reserves are scarce and civil areas predominate. There is acute scarcity of fuel and fodder, and most of the women have to walk almost 40 km to reach high altitude oak forests which are not under some kind of protection of villages enroute. Meanwhile they manage their few existing trees and grassland by opening the areas for lopping of coppicer shrubs every six months. Okhala, which has about 51 hectares of land in the civil, opens the area for cutting and lopping of branches only once in three years. Chir predominates in the reserve patch and khair sheesham, gudyal, bhimal in the civil. In the 20 hectare plot of Nakot in civil soyam land, there are less than a thousand trees of khair, sheesham, bel and semla, but the villagers have a chowkidar to guard them while also protecting their cultivation. He receives an annual phant of Rs. 60, with grain, from each family. Chanthi similiarly protects about 2000 trees of khair, sheesham and semla which are permitted to be lopped once a year when the area is opened for about 15 days. It is, however, kept open round the year for grass and this practice of keeping some grass 'bittas' for free foraging was found to be quite common in the villages surveyed. This convention, neccessitated by the prevailing biomass demands, (which have to be met somehow), would perhaps account for the vast denuded stretches which are a grim and common sight in Pratapnagar block.

Ghandialki and Sem village have a common area in the reserve forests of Narsinghtok, measuring about two and a half hectares with only 50 sheesham trees and 75 khair. The mahila mandal or womens village organisation had appointed a chowkidar who left because they were unable to collect the money for his pay. However this does not seem to have affected the protected status of the plot, since consensus for its care still exists amongst the women of both villages and their informal joint management continues to be respected by all.

Kotga talla malla, Jakhanikhal and Banali

These four villages have their forest areas in common. Chandni bitta is used by all for harvesting grass and twigs, and is free from controls, while Hatyur Khudgoda is a 30 acre patch, which the villagers have been jointly protecting, since the past 12 years, with the help of a paid chowkidar. The jungle is opened only twice a year, once for lopping of branches and once for grass.

Villages without any protected forests

Dandasari, Thala, Kalony Jakhwal gaon and Kudiyal gaon have no exclusive forest area of their own and have to share with others like Majaf, Silwal gaon, Pansut and Shukri, who all have their own forests in addition. The sahyoginis could not give much information beyond the statement that these areas are not governed by any rules of management. It can be assumed that these fall partly in the civil soyam no-man's lands, and partly in the patti high altitude forests which are now under increasing stress, not only because of general depletion of forest resources, but also because increasing protection in the vicinity of settlements would naturally lead to concomitant rise of biotic pressure on remote upland forests.

Baseli and Kouldhar: These two villages have no protection mechanisms, Kouldhar because of encroachments on common land and Baseli because they feel the land is too degraded, being without any trees and therefore not worth the investment involved in closure and watch and ward.

Village Protection of chir forests

Sahyogini Bichana came up with revealing facts regarding voluntary protection in the chir zone. These accounts indicate that villagers seem to have a pragmatic approach to the changing configurations of a fast decaying environment where beggars cannot be the choosers. The historic antipathy to chir- (highlighted by chipko leaders, as also by Meera Behn) seems to have been relegated to the past. For some time, one was under the impression that, with a few notable exceptions like Adwani, Sankri, areas adjoining New Tehri, by and large, chir forests were being protected by villagers only with the ultimate aim of phasing out the chir trees and allowing banj to take their place. However, Sahyogini Bichana has identified atleast seven gram sabhas where villagers are paying heavy sums, ranging from Rs. 600 to Rs. 1000 a month, to chowkidars who police the chir forests falling largely in the reserves. This protection, and the monthly household donations of Rs. 10-12 which the villagers are ready to bear, assumes even greater significance in view of the fact that these arrangements are being maintained even where the forest department is having its own chowkidars in the reserves. Some of the chir areas which have this dual system of chowkidari are Sirkhuli, Chamlyana, Chajoli and Sujjadgaon area of Byongi and Dhagunda and Kakari ki Patali and Sil area of Baunsari, Charka Bhedi, Patala, Sirkuli of Jamund and Semal belonging to Gramsabha Kyari, Kalapahar, Baghaghati and Musa ki Chauli belonging to Baunsari. While a 20 hectare plot in Deval was handed over to the department for plantation, when the latter withdrew their chowkidar, the villagers immediately resumed their watch and ward. Perhaps because of the size of the area and the density of the biomass, the villagers contribute Rs. 15 per head and pay Rs. 1000 per month to the chowkidar of Deval. This is primarily chir jungle.

Sahyogini Vimala also came up with detailed lists of chir forests being protected in the reserves, even though the villages listed were principally guarding large areas of prime broadleaf forests. Shukri, with the largest and oldest banj jungle in Vimal's beat, is protecting its chir forests in the reserves and this village watch and ward continues, despite the fact that the department has its own chowkidar in the chir jungle there. A similiar situation exists in other reserve forests being guarded by villages like Kharoli, Ramolgaon and Dang, Okhala I and II amongst others.

Conclusions

Information gathered by the sahyoginis indicates that, in the present environment of extreme deprivation which exists in this block, all forest species are being brought under some kind of management and direction, wherever a homogeneous settlement can be found having undisputed usage rights to a specified area, along with community will to act collectively for its protection. This community will may be actualised through a functioning panchayat or a local dedicated leader or it can be facilitated by pre-eminence of authority of elders. It can be set off by considered decisions, taken by informal groups, to end internicine quarrels over biomass extraction. These are just some of the factors which can lead to the kind of constructive management practices being described. The example each sets to the other also has a snowball demonstration effect, which is of critical importance for the spread of any kind of development programme in rural and backward areas.

Some interesting examples of successful co-operation between village groups and the forest department, as also NGOs, emerged from the survey undertaken by Guddi Chauhan. In Nautad namak tok the department had carried out a highly successful plantation about 20 years ago. It now has about 3000 trees of sandan, guryal, semal etc and natural regeneration is also prolific. The people continued to protect it after it was handed over by the department and because the site is believed to guard the village against soil erosion and landslides, green felling is still not permitted. In Pipalogi, an NGO has started a nursery and constructed a small pond or chali for harvesting water. This NGO has taken up a tree plantation program in the area being protected by the villagers and is accordingly utilising the services of the chowkidar appointed by them. They pay him Rs. 3000 annually. This is in addition to the collective phant of Rs. 3000 being paid to him by the villagers. The project has managed to elicit alot of public participation and the women regularly work on the project for daily wages. These are encouraging pointers towards strategies of protection which could be devised in tandem with the the local people.

Most villages in Pratapnagar block seemed to have adopted a system of annual meetings and resolutions which are passed by the village assembly. But these procedures are laid down and enforced mostly by men. It will be seen presently that the conditions of management are quite different in Bhilangana valley, where supervision is more informal and carried out almost entirely by women.

With an average of thirty households per village, and forty two villages paying , on an average, Rs. 60 annual phant per household, these villages are collectively incurring an annual expenditure of not less than Rs. 75,600 per annum for watch and ward. Over the past two decades, these 42 villages of Pratapnagar have contributed not less than Rs. 15,00,000 towards restoration of these prime Himalayan forests — a contribution which is neither acknowledged by the state nor officially registered by research studies, many of which, (despite recent encouraging trends to the contrary), continue to blame the people for degradation of the Himalayan forest wealth.

It is a simple ground truth, known to villagers, that the canopy of banj easily reestablishes itself in three years. If only this practice could be adopted, by consensus, in all the remaining oak forests which are under acute stress, but are not under any indigenous management system, we would still have a very fair chance of saving our extant oak forests which are dwindling largely because they are open to a whole patti or group of villages, and are therefore not under custodial charge of any particular settlement. How to achieve this consensus where the nexus and umbilical cord with community forests has been broken, is the million dollar question today for environmental planning in the Himalaya.

NOTES

1. Neeru Nanda : Tawang - *The land of Mon,* Vikas New Delhi, 1992.
2. Christoph Von Furer-Haimendorf Himalayan Barbary John Murray, London, 1955.
3. Personal communication from Mamta, Sahvogini Mahila Samakhya, Jakholi Block. Tehri.
4. *Bainyon ka Raibaar* : Pahla Sandesh - *Aurat ka Dookh* Doorsa Sandesh- Sharaab ka dhancha, Teesra Sandesh - Aurat aur Jungle. Mahila Samakhya Tehri Samooh : Mahila Samakhya Parivojna, Garhwal.
5. Kusum Rawat *Hamara Safal Paramparagat jungle prabandhan Mahila Samakhya,* Tehri (unpublished).
6. Aniruddha Brahmchari 'Experiences in forest management' in The Review, Vol. III No. 3, March 1995, New Concept Information System Pvt. Ltd, New Delhi.

CHAPTER 13

Eco-Management by Women's Groups

By and large, women's management practices are seen to be unconstrained, non- institutionalised, apparently unstructured and yet extremely effective. The stories of some villages give a fairly vivid picture of how the systems devised by women's groups work on the ground.

Over 32 settlements in the Balganga and Bhilangana valleys were found to have interesting protection mechanisms in both civil soyam and reserve forests. In the chir reserves most villages have a system of appointing chowkidars for six months, generally from July to November in order to supervise closures of the jungle for protection of grass, as also to manage the subsequent extraction. In local parlance this is called jagwal. Hence the reserves in these valleys are also policed under two separated systems *i.e.,* the jagwal of the community for grass and the departmental 'patrol' for the trees. Both seem to co-exist more or less peacefully.

The jagwals in this area are being completely managed by women. The women hold the annual meeting for closing the forest, they select the chowkidar, and meet whenever there is a serious conflict which has to be resolved. On occasions of inter-village conflict, they invariably take it upon themselves to patrol the forest and apprehend offenders. The fines system also appears to be less formalised as compared to Pratapnagar, perhaps due to the absence of male domination.

Once in a while, when conflict between village groups assumes serious proportions, the men occasionally step in. Often women themselves take initiative to elicit help of pradhans (past and present). Generally women's management practices appear to be quite informal. Each task is approached individually by sitting together and calibrating optimum solutions.

Control through magico-religious belief

Bhanede-Devaleshwara and Peda gaon collectively own an area which consists partly of banj forest and partly of grassland. The chir area is patrolled by the forest staff, while the grass areas are not under any management controls. However the banj jungle has been collectively 'offered' by the village women to their devata (Deity) — devata ko chadha rakha hai. Since the jungle no longer belongs to them, but to their God, it is quite safe and the trees have increased in size and girth. No woman will wield her dranti on these trees, and only the women who took the samkalp or oath before the devata are considered competent to lop them properly. Sahyogini Laxmi reported, however, that the younger village women are now showing an inclination to lop the branches, on the plea that they have not taken any oath and will not, therefore, incur the wrath of the devata!

In Riyunti village a one hectare banj jungle has been closed by a similiar process initiated solely by women. 'Shiv ke mandir mein chaval chadhana', is how they express it locally, since the pledge is taken by offering a handful of rice to the God while vowing to protect their oak forest.

Conflict Management

Kepars, Dharkot and Mandara: This story is interesting because it shows how traditional protection practices can also cease to be effective unless rejuvenated every so often. This would be specially so in the case of areas used collectively by more than two village groups.

These three villages had been protecting a beautiful oak forest in the reserves since the past fifty years in areas known as Burkhil, Dalpani and Mathidanda. Though villagers had themselves reared the trees and watched them gain in strength, with the passage of time, they became indifferent to their fate and started damaging the jungle through careless lopping and other malpractices. When the mahila samakhya started working in this area, the sanghas and sakhis of Kepars and Dharkot called a series of meetings, to which Mandara villagers were also invited. After a lot of deliberation, preceeded by some mutual recrimination, it was decided that each household would take turns in patrolling the jungle. This system did not work and thefts and indiscriminate lopping continued to take place. The sakhis got the villagers together once again and more meetings ensued. After three or four protracted sessions, it was finally decided that each village would appoint its own chowkidar and these three chowkidars would guard the jungle by monthly rotation. This system worked well for many years. In the initial period the chowkidars were selected with great care and the forest thrived. In 1996, however, fresh problems arose. Chowkidars became slack, while some women of Kepars and Dharkot made the mistake of stealing from the private protected oak patches of neighbouring Lasyalgaon with most unfortunate results. The Lasyalgaon people descended onto the protected oak forest of these three villages and wreaked vengeance by felling the standing oak trees. A major conflict ensued. Injuries were inflicted on either side in the resultant clashes. The real culprits remained unrepentant and defiant, while the Lasyalgaon people were uncontrollably bent upon taking revenge for the losses they had incurred. Finally, Sahyogini Laxmi collected the people in the forest where the young trees had been damaged and got both sides talking. They talked and ventilated their grievances for a long time, after which the Lasyalgaon people came around to the extent of promising to honour the protected forest of these three villages, provided they could control the women who were stealing banj loppings from their fields. It was finally decided that the offenders, if caught, would be punished by Lasyalgaon, though the fine of Rs. 25 fixed for each offence appears to be a trifle low, considering all the fuss and bother which had to be incurred due to these errant women!

Srikot Saud and Bhandari : These villages witnessed a historic conflict of considerable magnitude over distribution of grass. While the forest had been jointly protected since many years, a problem arose due to imbalances which had crept into the family structures—while many families retained the traditional pattern of agricultural households, others had reduced in size due to migration of the menfolk. Hence households with more hands for labour started appropriating the lions share of the grass, while women residing alone with small children sometimes could not manage to get even a headload. This naturally led to heart burning and quarrels, since all households pay the same amounts as phant.

After the inception of Mahila Samakhya, a covenient forum emerged for sorting out these problems. After much discussion, a system of one dranti per household was agreed upon. However, this was circumvented by some families who sent their menfolk or other women halfway up to the jungle to relieve their women of the load, and in this fashion, some people continued to take out more grass than others. The matter was again discussed and finally resolved with the help of one villager named Prem Singh who is fondly regarded now as a great friend. Prem Singh stations himself outside the jungle along with the chowkidars and two women volunteers. Each outgoing load is weighed and 50 kg allowed to be taken out free, while the excess is sold out at the rate of 50 n.p. per puli. At present, 90 drantis use the forest in this regulated manner and the daily sales range between Rs. 15-30. The money is deposited in the bank and controlled by two groups of women, one being in charge of drawing the money out, while the other checks the passbook and keeps the accounts. Each family gives one puli of grass per day to Prem Singh. This system of weighing the grass loads, which is in vogue in parts of Pratapnagar block (e.g. Dang-Ramolgaon), is also found in Beleshwara, Kangra and Chamiala of the Balganga valley.

Bhanedi and Turgana: These two village groups had serious differences because, though the forest area was traditionally subject to combined usage, there was perhaps more pressure on the forest which lay closer to Bhanedi and Paida. While Turgana continued to claim usage rights on that side, they refused to allow the other villages of Bhanedi and Paida into the jungle near their settlement. The conflict was finally resolved after protracted meetings, and that too, only after the area was divided and the two groups went their separate ways.

Nautar-Saroli (Jakholi Block): Sahyogini Mamta Nautiyal brought up the case of a plantation which the sangha and sakhis of Mahila Samakhya have been trying to raise in the reserve forest area of this village which has no civil soyam land available. At present the forest has only chir trees and the ground flora hardly yields any biomass. In 1996-97 the women got together and planted various broad leaf species like bamboo, kachnar, sandan, semla, aonla within the reserve and also closed it for protection. This step led to a lot of opposition from a section of the people, mostly men, who claimed that the women had no legal locus standi for planting trees in the reserve area, this being a prerogative of the forest department or the gram sabha. The conflict reached alarming proportions with Mamta being attacked and hit on the head. It seems that the opponents all belong to a separate settlement within the same gram sabha located across a small stream, and they used to originally use this area for grazing their cattle in winter when the stream dries up. With the recent construction of a bridge, however, they are now in a position to use it more frequently. As a result the Nautar women were being increasingly deprived of their original share of grass. According to them the other settlement has sufficient civil soyam land but almost all villagers have encroached upon it. Another interesting aspect which has emerged from Mamta's report is the involvement of local men and department staff in clandestine felling of chir. It seems the women had enclosed and started patrolling the area from where the local netas were prone to remove trees at night, get them sawn in a nearby gadhera (stream bed) and then transport the sawn wood to the plains. Having come up against very strong vested interests, the mahila sangha had to beat a tactical retreat. What hurt them the most perhaps is the fact that the village has one Prema Devi as one of the few female pradhans. (About one-third of the pradhans returned during the last hostly contested panchayat elections in Tehri were women.) The samakhya women scornfully dub them as purushon ki kathputli—mere puppets in the hands of men. This observation has been borne out also in Jakholi where Madhu Bhatt reports that a van panchayat has been formed in Makhet but the womens group which has actually been managing the area since the past five years was unaware of its formation. On further enquiry it turned out that the patwari had set the wheels in motion by contacting those very women, (obviously from influential families) who never co-operated with the mahila sangha in observing the rules and also never attended their meetings. It is quite likely, therefore, that in-depth studies might be able to trace a class-bias underlying the whole process and working of these formal and informal groups, with the few influential families being in command of the former and the vast majority of the poor comprising the latter.

Other Villages protecting forests

Unlike Pratapnagar, where largely single gramsabhas or settlements act as the unit for protection and management, quite a few village groups are seen to combine to protect fairly vast forests of broad leaved trees in the Bhilangana valley. From the duration of protection, such combinations appear to be quite recent and therefore it is possible to derive hope for similiar association elsewhere.

A 1200 hectare area, known as Pauni wala Patala or Banjyan, has been safeguarded by five villages since the past 50 years. The chowkidar of this huge banj jungle is paid Rs. 1500 per month. Similiarly, Silyara I and II and Chopriyan collectively manage an area only for grass, with an annual phant of Rs. 50. Likewise, Rajgaon and two other villages pay a phant of Rs. 60 for collective protection of grassland for a period of six months, for which the pay of the chowkidar has been fixed at Rs. 800 per month.

Kandarsyun has a protected banj and malhu (bauhinia vahili) area of about 200 hectares which is managed by the women themselves. Th chowkidar is also appointed by them. Sema has a beautiful 100 hectare deodar jungle being protected in the reserves since 45 years. Syura has been managing chir and banj jungles in the reserves since 50 years, while Padokha people reportedly regard a huge 1000 hectare banj area located quite far from their village, as their traditional forest which they have been actively guarding since over fifty years. Sendul, Camera, Saranshgaon are also guarding forests and grasslands largely through women's groups.

Villages without protected forests

Curiously, the list of villages without any protection mechanism in Bhilangana block is longer than that of Pratapnagar and it is certainly not exhaustive. Siludi and Gadara of Gadara gramsabha have open grazing in the civil lands. Camera saud and Sonargaon have no jagwal for grass because they have no alternative area to shift to. However, after the Mahila Samakhya campaign for planting trees, they have enclosed a portion for plantation. Cheli, Manikyanath and Anva have no jagwal for grass and their banj forest is not under any specific management, perhaps because the three separate villages have not yet found it neccessary to combine for this purpose. They do not operate any watch and ward in the reserves.

The differences between Bhilangana and Pratapnagar

The discerning reader would have been able to pinpoint a number of striking differences in the management of forest resources in both the blocks. Why is management in Pratapnagar dominated by males? Why is there a preponderance of protection by single settlements in Pratapnagar, whereas upto four and even five village groups are found to be cooperating for management of grasslands and forests in the Bhilangana block? In Pratapnagar the village chowkidars in the reserves seem to guard trees as well as grass, whereas in Bhilangana the jagwal in the reserves is only for grass. Lastly, there seem to be more villages extracting biomass from banj jungles without any formal or informal system of control in Bhilangana, whereas in Pratapnagar it would be difficult to find even a single oak wood which is not under some kind of regulatory management.

The answer to these questions perhaps lies in the relatively better state of the forests of Bhilanagana block both in terms of area and density. Bhilangana block has reported less lopping of trees for fuel and greater reliance on middle storey shrubs and thorny trees which have excellent calorific value. Perhaps this is due to the relatively better state of health of these forests which is borne out by statistics and even by an eye survey. Fisher et al also tended to conclude that indigenous management systems are a response to and correlated with, scarcity of resources.[1] Their hypothesis that people are unlikely to invest time and energy in organising access to a readily accessible source seems to be borne out by the comparative study of these two blocks. (Though it is contradicted by the presence of intricate norms for distribution of these resources existing in remote areas of relative plenty, like Gangi and Tawang, which have been studied and documented by the writer).[2]

The money economy plays a crucial role in Pratapnagar block, which is known for its big contractors, as also for illicit liquor business. Money cannot buy fuel, and in the conditions of near desperation which exist in that area, perhaps it is not surprising to find men playing a major role in the protection of surviving forests and also laying down heavy fines as punishment. In Bhilangana block, perhaps we see vestiges of a more ancient system, managed by women in times of plenty, with informal but clear norms. The Mahila Samakhya could also find a better environment in the Bhilangana block for creating the kind of forums which Laxmi Bisht and others have been successfully doing for conflict management and working towards co-operation between a sizeable number of discrete village settlements.

Needless to say, the polar types described here are not exactly replicated in the field. Even in Pratapnagar Sahyogini Guddi Chauhan reported a case of 3 or more villages uniting for protection, *i.e.* Dodag, Thapla and Dodalgaon who have been jointly appointing a chowkidar to protect a small jungle since the past 25 years. However, womens groups were playing a major role in this as in other villages of Naughar, Syalgi, Ghvera, Gad Kalyan and Paturi, all reported by Sahyogini Guddi Chauhan who was the only one to come up with reports of womens participation in protection in the villages falling within her juridiction in Pratapnagar block.

People's watch and ward vs Departmental watch and ward

It is interesting to see how both the people's and the sarkari programmes coexist. The department carries on its work in both civil soyam and reserves, but the chowkidars appointed by them are, in the words of the sahyoginis, only "nama matra"—only in name! The people keep their own chowkidars and simply ignore the government programme !

The department chowkidar is not subject to the close control and system of fines and accountability laid down by the people. Supervision by his senior officers is just not possible, given the poor infrastructure and set-up of the department. He seems to exist, therefore, in a kind of limbo, being sidelined by the villagers, with an apparently minimal role in protecting the jungle.

In terms of chowkidar salaries, both systems are at par only in some villages. The department chowkidars are paid roughly Rs. 900 per month in the reserves, but the plantation chowkidars used to be very badly paid, while the village chowkidars are generally paid Rs. 6000 per annum. Instances of chowkidars being paid upto Rs. 1500 per month have also been mentioned. In terms of regularity or periodicity of payment, both are at par, since the government pay packet arrives once in six month or sometimes after a year, while the communities make payment once in three or six months. However, the people's chowkidars enjoy an enhanced status by virtue of their power to impose fines, seize drantis, disgrace offenders and generally make their presence felt. In addition, many enjoy perks which vary with villages—some give them the fine money, most give cereal in addition to cash and almost all have a system of also giving chowkidars one puli of grass per dranti per day when the jungle is opened for a specified period.

The community chowkidars are also subject to stricter supervision and control. A government chowkidar is in a much more comfortable position in this respect. He cannot be removed after three years continuous service on the territorial side, whereas the villagers deliberate on appointments every year afresh in a meeting of the village assembly. Some even fine him if the jungle is found to have been seriously damaged.

Because of the element of payment by each household and active consensus for protection, the village chowkidar can bank upon the women's groups and the panchayat or village assembly to support him against habitual offenders. The department chowkidar, on the other hand, can only look for support from a forest guard who is harried by many demands and has too large a beat to be able to enforce effective patrolling.

The names of chir areas being protected in the reserves have been listed in some detail in the previous as well as the present chapter in the hope that forest officers may, at some opportune time, like to go into the details of how these two systems operate, interact and coexist. Are the forests under this dual system having better indices of health in terms of protection from overtapping, burning and felling by storms, as compared to overall averages for the chir zone in the district as a whole? The density and canopy cover of the chir forests, the bio-diversity and quantity of the middle and under-storey under dual protection systems would also merit some comparison with others under lone departmental protection.

Except for a few chir areas which are being guarded since 30 years, the majority appear to have come under voluntary protection only since the past 10-12 years. It would seem that community watch and ward of chir forests is a relatively recent phenomenon.

There is a long history of conflict between the department and the people, centering around firing the carpet of chir pine needles before the onset of the summer season, the best analysis of the problem having been made by Ramachandra Guha.[3] Till date, no really permanent and secure management procedures have been mutually agreed upon between the two parties, keeping in view their apparently opposed interests.

It is a pity that the department could not take advantage of these very active people's groups for safe custody of the chir forests which are particularly prone to fire. It would be interesting to have a detailed survey of areas devastated by fire in the recent past, when fires in the forest have really reached disproportionate levels. Investigations might possibly indicate that chir forests covered by people's protection mechanisms suffered less damage from fire than those without any such local arrangement. If it can be demonstratively proved that these management methods are actually effective in fire protection, the department could gain tremendously from this experience.

The villagers control extraction of grass and lopping in the chir forests, but the allocation of timber trees under the rights system (ramanna or chant), remains with the department and operates through the more conventional network of gram sabha pradhans, patwaris and forest guards. If this is changed to involve these user groups the possibilities of leakage and illicit felling might be minimised. Sawing and retrieval of trees felled in storms could also be handed over to these groups in the hope that illicit felling of green trees on this score would also be checked.

Participative management of chir forests on the lines of Joint Forest Management started by the forest department in West Bengal might prove beneficial for Tehri Garhwal in particular and the U.P. Himalayas in general, provided the department sits with the people and women's groups (such as those formed by Mahila Samakhya and other NGOs), and works out plans which will meet the felt needs and problems of the local people. At the time of writing, a world bank sponsored JFM project was in the pipeline, and some Forest Committees had reportedly been started. But the grass root involvement and village contacts of these committees have yet to be assessed. It is quite likely these will also end up as parrallel structures, existing on paper, while the indigenous management stuctures continue. That, however, is an optimistic scenario.

Fisher et al, in their perceptive study of similiar indigenous systems in Nepal, have very rightly pointed out that where reasonably effective indigenous management systems exist, external intervention is often counter-productive at worst and irrelevant at best. As mentioned in an earlier chapter, they found that even the appointment of externally sponsored forest watchers may lead to the breakdown of a local system of guarding the forest, as had happened in Thulo Siru Bari and an adjacent village.[4] Fisher et al have therefore concluded that negotiating an agreement which has no relevance to users is likely to be destructive, and a strategy of recognising the existence of these systems, and then effectively leaving them alone, might be the best one in some circumstances.

In the civil soyam, recently there has been a big thrust towards establishment of Van Panchayats in general with a stress on Mahila Van Panchayats where a woman is appointed as the sarpanch.

Details of the area covered are given in the table below:

Area		*No. of Van Panchayats*	*Total Area of Panchayats*	*Average (Hectares)*
Proposed	00 - 10	13	50.2	3.9
	10 - 20	9	114.4	12.7
	20 - 30	4	96.5	24.1
	30 - 40	4	142.6	35.7
	Above 40	2	133.0	66.5
Approved	00 - 10	66	291.1	4.4
	10 - 20	31	428.0	13.8
	20 - 30	9	201.2	22.4
	30 - 40	9	307.2	34.1
	Above 40	14	1349.8	96.4
Total		161	3114	31.4

Source: District Magistrate Tehri Garhwal, New Tehri 1997.

These figures show that during this decade van panchayats have been formed or are in the process of being formed in about 31 sq km of civil soyam lands which in all cover an approximate area of 545 sq km in the district. Hence, given the fragmented state of civil soyam lands and the pressures which they face, this initiative, by itself, seems unlikely to lead towards any radical change in the situation in which we find the hills of Tehri Garhwal today.

NOTES

1. R.J. Fisher et al ***The Management of Forest resources in Rural Development: A case study of Sindhu Palchok and Kabhre Palanchok district of Nepal,*** Nepal Australia Forstry Project, ICIMOD Joint case study (unpublished).
2. Neeru Nanda *Tawang—The Land of Mon,* Vikas 1982, see chapter 4 and 5 on graziers of Mukto circle for norms of distribution of grazing grounds.
3. Ramachandra Guha. *The Unquiet Woods,* Oxford 1989 pp. 50-55.
4. Fisher et al *op cit.*

CHAPTER 14

Future Trends

What promises does the future hold for the women of Tehri Garhwal? A series of new initiatives have been taken by the U.P. government since the start of this decade, particularly in the sphere of starting van panchayats in the civil soyam and joint forestry management in the reserves. These are laudable efforts and should be fruitful provided they are not sought to be operationalised through the same bureaucratic procedures, (based on mechanical fulfilment of targets imposed from above), by which earlier programmes for restoration of the forests were implemented. What we need most in the U.P. hills, perhaps, is a change in the outlook, orientation and management structures of the government departments dealing with the subject.

As government servants we need to change ourseives, the way we look at the people whom we are required to serve. This is what Meera Behn wanted. We do not need to change the people. They are already responding as best as they can to a very difficult situation by mustering the meagre resources available with them. The concept of catalytic agents, either in Govt. or urban based NGOs, is not just elitist—it is fallacious. Traditional wisdom is more ancient and rests on much firmer foundations than programmes planned and conceived by centralised bureaucracies and international aid agencies, which, howsoever well meant, are still far removed from, and far less effective than, peoples own efforts. What we need is a new respect for traditional wisdom and the common mans ability to manage his own environment. The Arabari experiment was based on this respect for the poor. Without this orientation (which has to permeate the entire forest department) mechanical policy formulations for JFM will not help.

In the U.P. Himalaya we need a recognition of the immense value of people's, especially women's, eco-management skills and experience along with recognition of the need for further extensive documentation of indigenous mangement practices for conservation of community property resources. The change has to come from within government. Our outlook, policies and structures have to change. Segmental approaches to the forest situation have to be replaced by a holistic approach embracing all categories of forest land and CPRs, whether in civil soyam or reserves. The forest department should be given full responsibility for all these lands—a policy initiative which is unlikely to come through, but which could be actualised by exploring alternatives like giving the forest officers concurrent jurisdiction under the land revenue act, so that they can atleast be effective in no-man's lands where the people's writ does not run and where protection may be sought for by villagers themselves. The present lopsided legal jurisdiction also mitigates against holistic approaches for forest mangement planning for village CPRs which fall in civil soyam as well as the reserves.

Within the forest department, major structural changes are required. Plantation and territorial divisions should be merged. The antique management structure whereby territorial divisions are expected to guard the forests and commercially exploit them, while the plantation divisions are expected to plant up new forests, is palpably ridiculous. A forest guard of the territorial division interacts with villagers only for protection and patrolling and felling, while the plantation division guard approaches them for closing areas for plantation. If the latter notices an offence being committed he is not expected to bother about it. The territorial guard may have a good plot in the reserves which he could motivate the villagers to close and protect, but then this does not fall within his charter of work since plantation targets and consequent funding in the reserves is severely limited.

It has been shown how ownership and exclusivity are the key factors in triggering off local protection practices. Surely the same factors would work for ensuring effective protection in forest guard beats ? But in the management structure of the forest department no single forest guard exclusively owns an identifiable area which he could possibly call his own. The areas of their beats are just too large to enable them to identify with even a segment. Even in these large sprawling beats, responsibilities are indeterminate and divided. I would venture to suggest that a simple policy intervention aimed at improving the management stucture of the forest department by making it more intensive, well-defined in terms of territories and sensitised to the importance of involving local user groups would be far more effective than the present well-meaning but confused thrusts towards van panchayats on the one hand and joint forestry management on the other. Both these initiatives so far seem to be labouring under the initial and basic defects of the plantation programmes—there is a top down approach, leadership initiatives are not emanating from within the U.P. forest department, the stress on target fulfilment is again leading to mechanical paper solutions and there is again a complete absence of an area approach. The policies of hop skip and jump continue.

The sensible and significant initiative of the Mahila Samakhya started by the Rajiv Gandhi government has been successful simply because it is not target oriented and because, both in belief and practice, it leaves a great deal of field initiative to women grass-root workers. The great contribution of the Mahila Samakhya lies precisely in this- that it has picked up, trained and empowered village women, given them the importance, opportunity and freedom to work out locally tailored solutions for locally perceived problems. It is this kind of facilitation and provision of local forums which is the need of the hour. The people of Garhwal do not need to be told what is to be done as much as they need to be extended a helping hand—just to able to get together, network a bit more than they can manage on their own, to be able to meet, talk, argue, listen and come up with their own plans for managing their micro-environment. Such networking would inevitably lead to more intensive interaction as equals with government officials. Villagers need to be helped to get together on forums which are non-political, where they can talk to government servants. It is a sad but inescapable fact that panchayati raj, while making great strides, has still many limitations in this regard. More and more non-political forums are needed where government departments, and particularly the forest department, meet and interact with the people and learn as they go along.

A somewhat similar approach had been adopted soon after independence under a different ideology influenced by Gandhiji's dream of self reliant villages. This scheme was launched under the once famous banner of community development which stressed self help and popular participation of rural people in the process of development of infrastructure and economic and social well-being in rural areas. Unfortunately, though the community development blocks were set up all over the country, they ended up being starved of funds since the community development programme could not withstand competition from more fashionable and lucrative schemes aimed at improving the economic status of individual farmers rather than the community as a whole. In the U.P. hills we still have some small water supply schemes and minor irrigation canals which were built in the hey day of community development with active co-operation of the community and are still being managed by the people themselves. The contrast between these humble but successful schemes and the world bank sponsored ambitious water supply schemes, not to mention hydrams, is quite interesting, since 60% of the water supply schemes set up with world bank assistance in Tehri Garhwal are non functional at any given point of time due to hurdles in the maintenance set up. Lessons can be learnt provided there are any takers.

Perhaps the greatest achievement of the Mahila Samakhya has been the fact that the whole documentation narrated in these chapters was completed by the sahyoginis themselves with minimal involvement of the district officials of the organisation. Many of the major conflicts narrated here were sorted out by the sahyoginis without making references to, or seeking help from, the district coordinator.

The importance of process against product has not yet seeped into the pysche of the revenue or the forest administration though it must be said to the credit of the latter that concrete efforts have been started by the Jalagam or Watershed Management Project in the Doon valley during the last two years. All these are hopeful signs—maybe straws in the wind. It is the whole emphasis on process which made the Arabari experiment a great success and it could be replicated within the state of West Bengal because the ideology and orientation of the district officials was sought to be changed through intensive and appropriate training. Unless similiar processes are adopted in the U.P. Himalaya, neither JFM nor van panchayats would deliver the results which are presently being expected from them, though definitely some results will emerge and we can only hope that they will be beneficial. Organisations like the Mahila Samakhya offer interesting possibilities for interaction with the department and the revenue officials on this score. As this book goes to the press, we are in the process of organising seminars and workshops of village women workers with district officials in order to generate awareness of these latest government ventures for ensuring people's participation in regenerating the forests of Tehri Garhwal. After a long time, we have found in Anand Singh Negi, Conservator Bhagirathi Circle, an IFS officer sufficiently senior and committed to take the programmes forward. So the story does not end here.

Perhaps a new, more hopeful, chapter has begun.

Or will history merely repeat itself ?

APPENDIX - I

List of Botanical standard and local names of common trees, shrubs and climbers

Sl.No.	*Local Name*	*Botanical Name*
Trees		
1.	-	Acacia molisimma
2.	Aonla	Emblica officinalis
3.	Ayar	Pieris ovalifolia
4.	Banj	Quercus incana
5.	Bel	Aegle Marmelos
6.	Ber	Zizphus giguba
7.	Bhimal	Grewia oppositifolia/ optiva
8.	Bluepine	Pinus excelsa
9.	Burans	Rhododendron arboreum
10.	Chir	Pinus roxburghi / pinus longifolia
11.	Dainkan	Melia Azadirach
12.	Deodar	Cedrus deodara
13.	Gudyal	Bauhinia variegata
14.	Haldu	Adina cordifolia
15.	Kachnar	Bauhinia acuminata
16.	Kaiphal	Myrica Sapida
17.	Khair	Acacia catechu
18.	Kharik	Celtis australis
19.	Kharshu	Quercus semicarpifolia
20.	Majnoo	Salix elegance
21.	Mohal	Pyrus lantana
22.	Pangar	Aesculus indica
23.	Poplar	Populus ciliata
24.	Sal	Shorea robusta
25.	Sandan	Ougeinia dalbergioides
26.	Shisham	Dalbergia sissoo
27.	Sirus	Albizzia procera
28.	Semla	Bauhinia purpurea
29.	Toon	Toona ciliata
30.	Utis	Alnus nepalensis
31.	Walnut	Juglan regia

Shrubs		
32.	Amelda	Rumex hastatus
33.	Dhawli	Woodfordia fructicosa
34.	Karaunda	Carissa spinarum
35.	Kingora	Berberis aristata
36.	Tungla	Rhus parviflora
Climbers		
37.	Maljhan (malhu)	Bauhinia vahlii
Weeds/Cactus		
38.	Kalabans	Eupotorim adenophorum
39.	Rambans	Agave
40.	Sura	Euphobia
Bamboo Cane		
41.	Bans	Dendrocalamus strictus
42.	Ringal	Arundinaria falcata

Bibliography

A. Government of India Publications

Eco-system of the Central Himalayas and the Planning Commission, Report of the Task Force for the study of Eco-development on the Himalayan Region, 1982.

Planning Commission, Report of the Working Group on Development of Hill Areas, 1985.

B. State Government Publications

Forest Department, *Report of the Kumaon Forest Grievances Committee*, 1922, Government Press, Allahabad.

Forest Department, *Working Plan for Uttarkashi Forest Division*, 1983 Tehri Garhwal State, Raturi, P.D.

Forest Department, *Working Plan for Garhwal Forest Division, 1930-31 to 1938-40*, Brahmawara R.N., Government Press, Allahabad.

Forest Department, *Tehri Garhwal Rajya Forest Manuals*. Saklani, I.D., Tehri Garhwal State : 1940.

Forest Department, *Working Plan for Tehri Forest Division, 1939-40 to 1969-70;* Khosla P.N., 1948.

Forest Department, *Forest Department, Annual Progress Report of Forest Administration, 1959*, U.P. Government Press, Allahabad.

Forest Department, *Report of the Kumaon Forest Fact Finding Committee, 1959*, Government Press, Lucknow.

Forest Department, *Centenary of Forest Administration in U.P. 1851-1961*, 1961, Lucknow.

Forest Department, *Working Plan of Tehri Forest Division, 1973-74 to 1982-83:* Verma V.P., 1974, Lucknow.

Forest Department, *Sattelite Imagery in U.P. Himalaya & Sivaliks*; Gupta P.N. Land Survey Directorate, Dehradun.

Forest Department, *Experts Committee Report on Felling of trees and maintenance of Environmental balance in the Himalayan Region of U.P.*, March, 1982.

Forest Department, *Integrated Watershed Rehabilitation and Development Project of South Bhagirathi for EEC assistance*. Land Survey Directorate, August, 1983.

Forest Department, *Bhagirathi Circle Ki Ropavani Ki Suchi*, 1986-87.

U.P. Government, *The Indian Forest Act with State Law* Beotra, of 1965, Govt. Press, Allahabad.

U.P. Government, *Garhwal Settlement Report, 1926-30,* Ibbotson I.W. Govt. Press, Allahabad.

U.P. Government, *Report on the 10th Settlement of Garhwal District*; Pauw E.K. 1895. Govt. Press, Allahabad.

U.P. Government, *U.P. Soil Conservation Act, 1954.*

U.P. Government, *U.P. Soil & Water Conservation Act, 1963.*

U.P. Government, *U.P. Private Forest Act, 1949.*

U.P. Statistics Department, *Forest Statistics 1960-61 to 1970-71.*

U.P. Statistics Department, *Forest Statistics of U.P., 1980-81.*

U.P. Statistics Department, *District Statistical Bulletin Tehri, 1985.*

U.P. Hill Development Secretariat, *Hill Development at a glance.*

C. Secondary Works: Books, Articles, Theses

Agarwala, Anil (ed) *The Price of Forests* Centre for Science and Environment, New Delhi, 1992.

Agarwal, Anil and Sunita Narain, *Towards Green Villages,* Centre for Science and Environment, New Delhi.

Amar Farooqui; *Colonial Forest Policy in Uttarakhand 1890-1928*; Kitab Publishing House, New Delhi, 1997.

Atkinson, E.T., *The Himalayan Gazetteer,* Reprint, Cosmo Publications, New Delhi 1981, Vol. I, Part II and Vol. III Part I.

Bajaj, Manjula., "Natural Regeneration Versus Afforestation: An Examination of the Potential for Investment in Natural Regeneration of Degraded Forests with Community Participation" in Anil Agarwal (ed). *The Price of Forests.* CSE New Delhi 1992.

Beadon Bryant, F., *Working Plan of Garhwal Forests Division, Central Circle; North West Provinces and Oudh,* 1896.

Buch, M.N., *The Forests of Madhya Pradesh*; Madhya Pradesh Madhyam Bhopal; P.O. Box No. 325; 1991;

Brereman G.D., *Hindus of the Himalayas* University of California Press, 1972.

Chambers, Robert and Saxena N.C., Tushaar Shah, *To the Hands of the Poor : Water and Trees,* Oxford and IBH Publishing Company Pvt. Ltd., 1989.

Chatterjee M and Ramble, *Himalayan Culture and Environment.* Indian Institute of Advanced Studies, Simla.

Chaturvedi M.C., *Water in Second India* Ford Foundation, New Delhi, 1972.

Chaturvedi, A.N., "Environmental Value of a Forest in Almora," Anil Agarwal (ed.); *The Price of Forests* CSE, New Delhi, 1992.

Chaturvedi, M.D., Chief Conservator of Forests Uttar Pradesh; *Role of Leaf Fodder in the United Provinces*; Government Press, Allahabad; 1948.

Dansey, E.P., *Working Plan for the Patli Dun Hill Forests, 1881.*

Dansey, E.P., *Working Plans for the Garhwal Forest Division 1879-80.*

Fisher, R.J., Hukum Bahadur Singh, Deepak Raj Pandey and Helmut Lang, *The Management of Forest Resources in Rural Development: A case study of Sindhu Palchok and Kabhre Palanchok district of Nepal,* Nepal Australia Forestry Project, ICIMOD Joint case study (unpublished).

Gadgil, Madhava, "Deforestation : Problems and Prospects" in *History of Forestry in India,* (ed). Ajay Singh Rawat, Indus Publishing Company, New Delhi, 1991.

Gadgil, Madhava, "Perverse Incentives for Industrialisation and its Impact on Forest Resources", in *The Price of Forests* (ed). Anil Agarwal; Centre for Science and Environment, New Delhi 1992.

Guha, Ramachandra, *The Unquiet Woods; Ecological Change and Peasant Resistance in the Himalaya;* Oxford University Press; Delhi; 1989.

Hearle, N.C., *Working Plan for the Ganges Working Circle, Ganges Forest Division, Central Circle, NWP and Oudh, Allahbad; 1888.*

Himalayas Sewa Sangh, *People and the Forests,* Rishikesh, 1977.

Indian Institute of Public Opinion, *Tree Survival Survey of five States* 1987 (unpublished).

Ives, J.D. 1985. "Mountain Malaise", in *Integrated Mountain Development,* (edited) T.V. Singh and J. Kaur.

Jayal & Bandopadhyaya J., *Indian Environment Crises and Responses,* New Delhi, 1984.

Jodha, N.S., *Common Property Resources and Rural Poor in Dry Regions of Indian,* Economic & Political Weekly 21, 1986.

Joshi S.C., *Rural Development in the Himalayas : Problems and Prospects,* Gyanodoya Prakashan, Nainital.

Kaur Jagdish, *Tourism and Pilgrimages in Eastern Garhwal Himalayas,* English Book Store, New Delhi, 1983.

Malhotra, K.C. & Mark Poffenberger, (edited) *Forest Regeneration through Community Protection: The West Bengal Experience,* West Bengal Forest Department, 1989.

Mahila Samakhya *Bainyon ka Raibar Jungle aur aurat, Bainyon ka dukh* Tehri Garhwal, Sector 2G 31 near Collectorate, New Tehri.

Mason, Philip, *A Shaft of Sunlight,* Vikas Publishing House, printed in Great Britain, 1978.

Mazumdar N.D., *Himalayan Polyandry* : Bombay Asia Publishing House, 1962.

Mehta, Manjari, *Cash crops and the changing context of womens work and status: A case study from Tehri Garhwal* (ICIMOD Discussion paper series Jan 93).

Mira Behn: *A Centenary Tribute; The Intach Environmental Series,* New Delhi.

Mira Behn: (Madeleine Slade;) *The Spirits Pilgrimage,* Longmans, Great Britain; 1960.

Mishra, Anupa, Satyendra Tripathi; ***Chipko Movement; Peoples Action for Development and Justice***, December 1978.

Nanda, Neeru, *"Local Mangement of Community Property Resources,"* Anil Agarwal (ed.) *The Price of Forests*, CSE, 1992.

Nanda, Neeru, *Tawang : The Land of Mon*, Vikas Publishers, New Delhi, 1982.

Nanda, Neeru, "Who is destroying the Himalayan forests" in Chatterjee and Ramble (ed.) *Himalayan Culture and Environment*, Indian Institute of Advanced Studies, Simla.

Nanda, Neeru, *Afforestation Strategies in the Bhagirathi Catchment*, Study conducted for National Centre for Human Settlements and Environment, E5/A, Girish Kuj, Arera Colony, Bhopal, p.58.

Nobu Masa, *The One Straw Revolution* Nataraja Publishers, Dehradun.

Osmaston A.E., *Working Plan for the Garhwal Forest Division, United Provinces, 1921-22 to 1930-31*; Government Press Allahabad; 1921.

Pande, D.C. Forest Management, in *Rural Development in the Himalayas*, (ed) S.C. Joshi, Nainital.

Pant, G.B., *The Forest Problem of Kumaon* (ed) Ajay Rawat reprint Gyanoday Prakashan, Nainital.

Pauw, E.K.; *Report on the Tenth Settlement of the Garhwal District*; Allahabad Government Press; 1896.

Pearson, G.F., *Sub-Himalayan forests of Kumaon and Garhwal, Report on the Bhagirathi Valley Forests*, in Selections from Records of the Government of the North West Provinces, 2nd series, Vol. 2, Allahabad 1869

Poffenberger, Mark & Betsy McGean, (ed.) *Village Voices Forest Choices, Joint Forestry Management in India*, Oxford University Press, 1996.

Programme Evaluation Organisation, Planning Commission, *Evaluation Report on Social Forestry Programme.* New Delhi 1987.

Rawat A.S., *Garhwal Himalaya; A Historical Survery (Political and Adminisrative History of Garhwal 1815-1947)* Eastern Book Linkers Delhi 1983.

Rawat, Ajay S., Joshi, D.P., Forest Policy for the Regeneration of the Himalayan Environment in *Environmental Regeneration in the Himalaya Concepts and Strategies;* (Ed.) J.S. Singh.

Reiger, H.S. 1981. 'Man Versus Mountain', in *The Himalayas : Aspects of Change.* (ed.) J. Lall and Modie.

Richard St. Barbe Baker; *My Life My Trees.*

Rui Gomes Pereira, *Goa: Gaunkari, the old village associations* edited and published by A. Gomes Pereira Panaji 1981.

Saxena, N.C. *Farm Forestry in North West India* Ford Foundation, New Delhi, November 1990.

Shashikant, "Privatisation of Common Property Resources, case of Tree Patta Scheme in U.P" in *The Price of Forests*, (ed.) Anil Agarwal, Centre for Science and Environment, New Delhi, 1992.

Shreshta K.B., *Community Forestry in Nepal, An Overview of Conflicts,* Discussion paper series International Centre for Integrated Mountain Development, 1996.

Singh J.S., et.al., *Environmental Regeneration in the Himalayas -concepts and Strategies*: Gyanodoya Prakashan, Nainital, 1985.

Syed Ahmad Pasha, *CPRs and Rural Poor. A micro level analysis* Economic and Political Weekly 14 Nov. 1992.

Thomas Weber, *Hugging the Trees* Penguin; 1989.

Tucker R.P., *The British Colonial System and Forest of Western Himalayas,* Duke Press Policy Studies, Durham.

Walton, H.G., *British Garhwal: a Gazetter,* Allahabad; 1910; Reprint Nataraj Publishers, Dehradun 1989.

Webber, T.W., *The Forests of Upper India* London 1902.

Glossary

Bhabbar	– Eulatiopsis binata; grass which can be used for pulp and also ropes
Bitta	– Plot of grassland
Chowkidars	– Guards appointed for watch and ward
Crore	– 10,000,000
Chaini	– Cattle herder's camp usually located in the forest or uplands
Chak	– Enclosure of grass
Civil soyam	– Revenue land adjoining habitation which is under common usage of villagers for grazing, collection of fuel and grass, extention of cultivation and also for setting up common facilities. Pattas or licences under this land can be issued to individuals on a system of lease. The patwari is the field level official who controls all revenue common land i.e. civil soyam as also private land in so far as he maintains all the land records.
Chipko	– Literally to hug or hold. Name given to the famous movement for protection of broad-leaved forests which started in Reni village in the Alaknanda valley in 1975 when a group of village women prevented contractors from felling a coup which they had purchased in public auction. This movement resulted finally in a ban on commercial felling in the UP Himalayas above an altitude of 1000 mtrs.
Compartment	– Unit of forest management in the reserves. One forest development block is divided into several compartments which are allocated between forest guard beats.

Desi	– Local or indigenous
Dranti	– Sickle; term used for number of women, armed with sickles, entering a forest
Devata	– Temple Deity; village god
Gram sabha/ gaon sabha	– Village council comprising all house holds in the village
Gujjar	– Migratory tribesmen, from outside Garhwal, who bring huge herds of buffaloes for seasonal grazing in the highland pastures on payment of grazing fees realised by the forest department
Gadhera/Rauli	– Mountain Stream
Jagwal	– Village system of management of grassland usually by women, involves about 6 months of voluntary closure of a given area.
JFM	– Joint Forestry Management
Kheti	– Cultivation
Lakh	– 100,000
Malguzari	– Land revenue levied by the state on cropped lands in British period
Mahila Samakhya	– Grass roots organisation for women's empowerment set up under a special project of Ministry of Human Resource Development in 1987
Mahila Mandals	– Women's groups organised usually at the village level and often formally registered with the Registrar of Societies. Mahila Mandals of Pratapnagar in particular played an important role in the anti-liquor movement that swept the block in the nineteen seventies
Neta	– Political leaders
Nali	– Measure of land, 20 nalis is equal to one acre
Panchayat	– Village council, lowest form of local government, consists of elected members headed by a chairman
Phant	– Refers to a flat contribution paid equally by all households in a village or group of

	villages for maintaining or guarding a common property resource, usually forest land or a plot of grassland
Patta	– Literal meaning is contract, refers to leasing of government land/common land on specific terms, also means title deed to land. (May also see civil soyam above)
Patwari	– Keeper of village land records
Patti	– A group of villages in the native system under the Raja of Tehri Garhwal
Puli	– A fistful of grass
Rabi	– The winter cropping season
Ramanna	– Each household's right to timber from reserves as allocated in forest department records
Sarpanch	– Chairman of panchayat or elected village body
Sarkar	– Government
Sahyogini	– Woman worker with an average of 10 villages in her jurisdiction in the Mahila Samakhya organisation
Sakhi	– Literally friend (feminine) designation applied to mahila samakhya, women organisation workers in a village
Sangh	– Women's group in a village
Samiti	– Committee
Zamindari	– System of land ownership by non-cultivators, which has now been legally abolished
Van Panchayat	– An elected village level body constituted for management of a specified forest (van) area generally carved out of civil soyam land. In 1919 the British government carved out remote tracts of thick forest in British Garhwal and Kumaon which were taken out of the Government notified forest tracts and reverted to the people in the form of van panchayats.

Index